Reptiles and Amphibians of the Southern Pine Woods

UNIVERSITY PRESS OF FLORIDA

Florida A&M University, Tallahassee
Florida Atlantic University, Boca Raton
Florida Gulf Coast University, Ft. Myers
Florida International University, Miami
Florida State University, Tallahassee
New College of Florida, Sarasota
University of Central Florida, Orlando
University of Florida, Gainesville
University of North Florida, Jacksonville
University of South Florida, Tampa
University of West Florida, Pensacola

Reptiles and Amphibians of the Southern Pine Woods

Steven B. Reichling

University Press of Florida
Gainesville/Tallahassee/Tampa/Boca Raton
Pensacola/Orlando/Miami/Jacksonville/Ft. Myers/Sarasota

13 12 11 10 09 08   6 5 4 3 2 1

Funding to assist in publication of this book was generously provided
by The Center for North American Herpetology, Lawrence, Kansas.

Library of Congress Cataloging-in-Publication Data
Reichling, Steven B., 1957–
Reptiles and amphibians of the southern pine woods / Steven B.
Reichling.
p. cm.
Includes bibliographical references and index.
ISBN 978-0-8130-3250-4 (alk. paper)
    1. Reptiles—Ecology—Southern States. 2. Amphibians—Ecology—
Southern States. 3. Pine—Ecology—Southern States. I. Title.
QL653.S67R45 2008
597.90975—dc22   2008011057

The University Press of Florida is the scholarly publishing agency
for the State University System of Florida, comprising Florida A&M
University, Florida Atlantic University, Florida Gulf Coast University,
Florida International University, Florida State University, New College of Florida, University of Central Florida, University of Florida,
University of North Florida, University of South Florida, and University of West Florida.

University Press of Florida
15 Northwest 15th Street
Gainesville, FL 32611-2079
http://www.upf.com

Dedicated to Dr. Roger Conant (1909–2004),
the first herpetologist to marry a love of pine woods aesthetics
to the quest for scientific study of its herpetofauna.

/12

Contents

Distribution Maps

Color Plates

Preface

I contend that the rapid and ongoing evaporation of the various pine woods ecosystems of the United States—habitats treated here under the heading "southern pine woods"—along with their endemic herpetofauna, is the most serious and urgent conservation issue facing North American herpetologists. Certainly there are other concerns of broad scope that justifiably compete with this issue for the attention and resources of dedicated individuals and organizations. There are other regional or biogeographic hot spots for herpetological conservation, such as the water allocation controversies affecting thorn forests of the lower Rio Grande Valley of Texas, or the disassembly of the Mojave Desert as a cohesive biozone due to mounting pressure from a rapidly growing concentration of humans. There are taxonomically framed issues cast over a larger geographic scale, such as the general decline of many riverine emydid turtle species or pandemic amphibian declines. Other pressing concerns can be viewed from either of these two perspectives, an example being the tenuous status of many troglodytic eurycid salamanders in the limestone outcroppings and escarpments of south central Texas. A herpetocentric conservation biologist can find much to be concerned with in these times, and books can be written about each of these topics. The reason I have come to write this particular book is that none of these troubled faunas are so critically threatened, nor so interesting in my opinion, as the community of reptile and amphibian species characteristic of the once grand southern pine woods, a habitat now so rare in its true and natural form that I find it increasingly common for biologists to be unable to distinguish between the pristine condition and the perverted version that now impersonates this assemblage throughout much of the rural South.

The overall region encompassed by this book corresponds closely to the historic range of longleaf pine. When Europeans first arrived on this continent, forests dominated by longleaf covered the majority of dry terrain along the Atlantic and Gulf Coastal plains. From North Carolina south to include north central Florida, it extended westward virtually unbroken until reaching the Mississippi River Delta in southern Louisiana. It resumed with a distinctive fauna and flora in central Louisiana, finally terminating at the transition with the central plains in east Texas (Boyer 1979). At that time, the longleaf forests blanketed approximately 92 million acres (Frost 1993) and undoubtedly harbored populations of the herpetofauna highlighted by this book in far greater abundance and geographic span than we see today. This once great forest ecosystem has been reduced to approximately 5 percent of its former size and fractured into a series of scattered fragments totaling slightly more than 3 million acres (Outcalt and Outcalt 1994). Expanding the geographic scope of this book somewhat is my decision to include other habitat associations in the southeastern United States under the mantle of "southern pine woods," such as the slash pine woodlands of central and southern Florida, the unique pine rocklands of extreme southern Florida and the Keys, and forests with mixed associations of longleaf, shortleaf, and sand pine embedded throughout the broad geographic span where longleaf once reigned supreme. All these closely interrelated pine-dominated habitats share the basic characteristics of true longleaf forest—sandy substrate and open, sunny floor—with no evidence that their inclusion softens the focus of the book.

The security of a herpetofauna can be evaluated according to several criteria. Foremost is the number of endemic species. Stated more pointedly, if a particular ecosystem were eliminated, how many unique life forms would be forever lost? Taking this list of vulnerable taxa under consideration, an assessment should determine the population status and degree of threat that each one faces. Are many of the life forms literally teetering on the edge of extinction, or are they simply less common than they once were? Are the factors causing the decline multitudinous and driven by powerful economic or social interests, or would a simple tweaking of some minor and regional activity eliminate the problem? Finally, when prioritizing a conservation issue in comparison to others, one must note the level of recognition of the problem and acceptance that indeed it is a problem by the conservation community and the public at large. Without broad support, genius and innovation born from synergistic partnerships, social and political clout to implement powerful ideas, and the resources that support them—not the least of which is money—will always be insufficient. This book is in part an attempt to garner such support for the particular herpetofauna of the southern pinelands.

By the terms of their inclusion in this book, each of the 26 taxa encompassed could disappear as elements of the wild fauna of the United States if the southern pine forests, as a functioning ecosystem, ceased to exist.

Unfortunately, this scenario has crept alarmingly close to reality in many portions of the distribution, and continuing trends elsewhere are unsettling. Of course, the endemic fauna of any biologically delineated area would be lost with the collapse of the underlying support mechanisms, but for no other large ecosystem in the United States is this so tangible a possibility for the herpetofauna as it is for the southern pine woods. Many of the most distinctive and characteristic reptiles and amphibians of the southeastern United States would be lost if the pineland communities disassembled, and examples of this very event occurring within decades are recounted repeatedly in this book.

Acknowledgments

I can trace the beginning of this book to a family vacation to Florida in 1970, when I was just 12 years old. We spent a week on Cedar Key. While my mother and father relaxed at the rustic inn and my brother and sister fought off boredom in the rural setting, devoid of beaches or playmates, I reveled in my first exposure to the dry, scrubby pine woods of the Deep South. Already determined to become a herpetologist, I spent every spare moment exploring the flatwoods and savannas that surrounded the town. The terrain looked so exotic to me, filled with odd and unfamiliar plants. I trekked through this landscape in a perpetual state of anticipation, thrilled by the prospect that I might at any moment encounter one of the incredible reptiles or amphibians known to be in these woods, species I had marveled at in the pages of my library of reptile and amphibian books. For sparking my lifelong interest in the herpetofauna of the southern pine woods, I credit my mother, Norma, and my father, George H. Reichling.

In addition to widely available facts about this group of species, which I have done no more than assemble under one cover, I have included many personal insights and perspectives gained from firsthand excursions to the pine woods. Most of these experiences were shared with companions who made the times more memorable and productive. Many trips to Louisiana were in the company of friends from the Memphis and Audubon zoos, namely Chris Baker, Dino Ferri, Amy Davis, Chris Tabaka, John Ouellette, Mike Wines, Michelle Keck, Justin Collins, Meghan Carr, and Kim Bailey. My guide to the pine rocklands of the Florida Keys was Robert Ehrig, who showed me corners of pristine habitat tucked away so well that I would never have found them on my own. In southern Mississippi, Mike Sisson took me to hidden sites in Harrison County. Dustin Smith, of the Lowry Park Zoo in Tampa, tolerated horrible weather so I could spend productive time in Polk County, Florida. Karen and Ricky Maestri are always intertwined in my memories of trips to southern Alabama. Joey Romanelli made St. Vincent Island conveniently accessible.

I have spent more time in Louisiana than in any other compartment of pineland habitat. Some of my trips to that region spanned many days in succession, and without the kindness of several key people, these visits would have been more difficult. The fine people at the Preheat deer camp, and in particular, John Girourd, were a real blessing. Mr. Pop Willard opened his home to me and treated me like a family member. The staff of International Paper Company's Bodcaw Unit, Charlie Colvin, David Whitehouse, Tommy Smith, and Paul Durfield, were instrumental in my fieldwork on land they managed.

Ray Semlitsch provided encouragement for this book during its beginning stages, and read some early drafts. My thanks go out to Linda LaClaire of the U.S. Fish and Wildlife Service for allowing me the privilege of participating in the Dusky Gopher Frog recovery program. Gary Lester was chief among the staff of the Louisiana Natural Heritage Program to grant me permission to conduct research in Louisiana pine ridge tracts over many years. Carrie Sekerak, U.S. Forest Service, was very helpful during activities involving Striped Newts in Ocala National Forest. Craig Rudolph, Doug Elrod, and Jeff Whitt were my partners during several years of research in Louisiana. I am very grateful to Lucinda Treadwell for her expert copyediting of the manuscript.

Researching and writing a book, for me at least, is an undertaking that tends to absorb and distract, but fortunately my employer, the Memphis Zoological Society, is staffed by people who value such projects and have always given me the freedom to pursue such things. At the Memphis Zoo, writing a book becomes just another aspect of my job, and is supported. I have been fortunate to work with many talented people at the Memphis Zoo. Some of them have worked with me directly, trying to devise the best ways to maintain and reproduce pine woods herpetofauna in captivity. Chief among these fellow zookeepers are Dan MacDonald, Robb Hill, Charles Beck, Danny Tennyson, George Heinrich, Mike Wines, and Michelle Keck. Chris Baker deserves special mention for generously sharing his computer expertise while preparing the range maps. I thank all my colleagues at the zoo, and hope that this book in some way reflects favorably on the place where we devote ourselves.

I am, at my best moments, only a mediocre photographer. I am most fortunate that people who are extremely talented in this art form were kind enough to let me share their superb work. The works of Suzanne and Joseph Collins shine in the pages that follow. Barry Mansell is another genius in the art of composing reptiles and amphibians in a way that shows their beauty to maximum effect. I am happy to acknowledge Robert Ehrig and his daughter Natalie Ehrig, who provided an image of the Key Ringneck Snake, one of the least photographed species of U.S. herpetofauna. It would be hard to overstate the value of the contributions of these individuals in the overall outcome of this book, as I could never approach the quality of their results, and I know the reader will enjoy seeing these beautiful images as much as I do.

Introduction

Defining the limits of a cohesive southern pine woods herpetofauna is not a simple task. Unlike a state field guide, the boundaries of a habitat are not predefined. More problematic is the fact that very few species of reptiles or amphibians are absolutely never found outside a landscape dominated by the principal defining elements of sandy soils and pine-dominated overstory. Some species, while certainly very typical of the southern pine woods, also exploit other types of ecosystems, such as the Pigmy Rattlesnake, *Sistrurus miliarius*, which is a characteristic resident of both pinelands and swampy situations, and Coral Snakes, *Micrurus* spp., which always seem present in ridge and flatwoods, yet are also frequently encountered in hammocks. Other species are completely restricted to the southern pine woods in parts of their range but more broadly distributed elsewhere, such as the Eastern Indigo Snake, *Drymarchon corais couperi*. Are these appropriate examples of pine woods reptiles and amphibians? The question I asked when considering each taxon for inclusion in this book was, "If the southern pine woods disappeared as a functioning ecosystem, would this organism persist somewhere within its distribution?" For the Pigmy Rattlesnake, the answer is yes; in the bottomlands and along river floodplains in Mississippi and Alabama, in swampy coastal areas of Florida, and elsewhere, the species would survive even having lost an important part of its distribution. The answer for the Eastern Coral Snake, *Micrurus fulvius*, would also be yes, it would survive. In the hammocks and backyards of south Florida, the Coral Snake would still be seen. Even the Eastern Indigo Snake, *Drymarchon corais couperi*, a federally listed Endangered Species, apparently extinct in Louisiana, Mississippi, and Alabama as a result of changes wrought on the dry longleaf pineland habitats there and now restricted in Georgia to this association, includes more generalist populations in southern Florida that are surviving in habitat quite different from the southern pine woods. The 16 full species and 10 subspecific races of herpetofauna defined by the scope of this book were so designated by virtue of their restriction to habitats that were at least historically characterized by a dominance of longleaf (*Pinus palustris*), slash (*P. elliotii*), shortleaf (*P. echinata*), or sand pine (*P. clausa*) as the overstory tree. While a few of these taxa are on occasion found in situations other than true pine wood habitats, this is an unusual occurrence everywhere in their range, and such individuals are generally in close proximity to pine woods where they likely spend time also. This is not a book about pine trees, but the same natural forces that led to *Pinus* holding sway over other trees are the very same forces that shaped a remarkably distinctive herpetofauna.

Considering its poor state of health, it is alarming to consider just how large and irreplaceable is the herpetofauna of the southern pine woods. Not only has the habitat nurtured an unusually long list of endemics, but at a surprisingly high taxonomic level as well. Only five of the taxa are subspecies of otherwise common and wide-ranging species: *Terrapene carolina major*, *Coluber constrictor helvigularis*, *C. c. etheridgei*, *Lampropeltis getula meansi*, and *Diadophis punctatus acricus*.

This fauna is an aggregation of some of the rarest reptiles and amphibians in the eastern half of the United States. The number of taxa with distributions restricted to areas so constrained that they can be concisely described by listing county names—a subjective dozen or so—is an ominous sign. Among the herpetofauna of the eastern United States, only a list of Appalachian salamanders would be a contender for more numerous minutely contained distributions. However, while the plethodontid salamanders of Appalachia have evolved and speciated on a very local scale as a result of abrupt transitions between contrasting physical characteristics of the landscape, most pine woods residents now occupy small ranges because of recent changes wrought by human activity over the broad landscape. The Appalachian salamander populations, while certainly fragile and notwithstanding a few exceptions, are as common and widespread today as they were when the European explorers first encountered them in the virgin forests. The same cannot be said for the reptiles and amphibians of the southern pine woods. Most species are disappearing or becoming scarce throughout their ranges. In some cases, such as the kingsnakes of the eastern Apalachicola lowlands in the Florida panhandle, the decline in abundance and areas inhabited has been so rapid that it was perceived within the brief time span of a single individual's field study of them.

It is amazing that an ecological disaster of the scale of this nearly total annihilation of the longleaf pine ecosystem is an esoteric conservation topic. Some, such as longleaf chronicler Lawrence S. Earley (2004), contend that the issue of longleaf conservation is well appreciated. I agree that among biologists who base their work in the South, the plight of the pine woods is never far from their thoughts. However, I have found interest in the subject to wane rapidly as one turns to those focused on other regions. Among laypersons, even those living on the Atlantic and Gulf coastal plains, the predominant response to comments made about the value of the pine woods, the distinctions between ecosystems, and the critically poor status of natural areas is one of mild interest to a subject never previously considered. While I don't hold much hope, my fondest wish is that this book will help advance the cause of pine woods ecosystem conservation by recruiting new supporters.

Distribution maps

Maps showing the distribution of terrestrial animals often come with the caveat that the area indicated is only a generalized expression of where that particular species might be found. In other words, anywhere within the shaded area, the species may occur if the proper habitat is present, but since habitats are discontinuous it is incorrect to assume that the animal in question lives throughout the highlighted area. For species that live in widespread habitats, disturbed sites such as agricultural land, for example, or those that have adapted to a wide variety of situations, these maps may be an accurate and concise depiction of where the species is found. In contrast, the southern pine woods have become so shrunken and dismembered that such generalized distribution mapping is misleading. Take the example of the Southern Hognose Snake. Tuberville et al. (2000) presented a distribution map for this species that dramatically illustrated the false impression a generalized map creates when addressing a rare habitat endemic. Southern Hognose Snakes occur throughout the southeastern gulf coastal plain, from North Carolina through northern Florida, westward to southeast Mississippi, and range maps depict this by shading the entire area. Tuberville and colleagues prepared two maps, one showing the historical distribution of the species by shading the overall region roughly defined by the scattered collection sites. The second was based solely on recent observations; a record for a particular county justified adding only that specific county to the map, so the distribution was not illustrated for any county without recent documentation. The generalized map gave the impression that the Southern Hognose Snake is ubiquitous throughout the southern gulf coastal plain, which it is not. The more constrained map revealed a highly fragmented distribution composed of disjunct populations, and not coincidentally approximating the occurrence of significant areas of pine ridge. I believe the second approach is more informative and accurate for all of the reptiles and amphibians coming under the scope of this book, but I have used a different methodology to attain a similar result to Tuberville et al.

My approach has been to first construct a detailed map of the significant pine woods sites within the nine state region. For some places, such as Eglin Air Force Base or Big Thicket National Monument, it is common knowledge that large areas of superb pine habitat are preserved. I have amassed information regarding more obscure sites in a variety of ways: consulting databases already compiled by agencies concerned with the preservation of longleaf and sandhill habitats, looking over topographic maps in selected regions where undeveloped upland sites are usually dominated by such habitat, and, primarily, personally investigating word-of-mouth reports and my own hunches. I have traveled widely while researching this book, always looking for just one more sandy road leading to a pine flatwood that I had previously overlooked. The result is a map of ecologically significant pine woods, from which the distribution maps of each species were derived.

Since I was interested only in pine woods that support a distinctive herpetofauna, the maps show only significant sites of occurrence. To be deemed "significant," an area needed to be both large enough and old enough to support its characteristic ecosystem. A few tracts of old growth longleaf, while very beautiful, were so small and isolated that they did not qualify as a place where pine woods herpetofauna could be found. These spots function more like paintings or living photographs—effective at showing what the southern pine woods used to look like, but no longer able to function as an ecological community. Conversely, many areas throughout the southeast are being replanted with longleaf in an attempt to re-establish pine woods where they once dominated. These efforts encourage optimism, and perhaps some of these young forests will grow large and stable enough to support a natural flora and endemic fauna. For the time being, however, they are works in progress and do not contain the reptile and amphibian species they might appear to, and they are not included in the maps for this book.

Once I had produced a base map of significant sites, I composed each species map by highlighting those areas falling within the generalized distributions that have been presented by other publications for the species. In this way, I hoped to correct the deceptiveness of more general illustrations of range given that the habitat required by these animals is so heterogeneously scattered within a matrix of unsuitable land. The maps that resulted from this process present a depressing picture, but I believe it is an accurate one, indicating that this fauna has been relegated to surviving in widely scattered sites, many being quite small and vulnerable to quick annihilation by human modifications. The error inherent in my method is that I have undoubtedly overlooked or mistakenly dismissed some tracts where a particular species does indeed occur, a problem not as likely to flaw distribution maps that cast a broader generalization of a species' range. With the subject of my maps being pine woods inhabitants, my method seems sound because suitable habitat is so spotty in occurrence. For example, a range map depicting the Marbled Salamander, *Ambystoma opacum*, as occurring throughout the entirety of southern Illinois would not be misleading to an outdoor enthusiast wanting to learn about the amphibians in that region, because even though they prefer certain types of situations, Marbled Salamanders can and do inhabit forests, fields, vacant lots, and backyards in this area. However, if I depicted the distribution of Mabee's Salamander, *Ambystoma mabeei*, as occurring continuously from the southeastern corner of North Carolina and extending across the whole of the South Carolina coastal plain, I would be including significant acreage of agricultural land, towns, and hardwood forest where the species would never be encountered. Even limiting my map for *A. mabeei* to areas dominated by pine trees would paint an inaccurately large picture of the territory it inhabits, owing to the loss of extensive mature pine associations and their conversion to plantation. It is sad but true that when addressing pine woods herpetofauna, range maps that run the risk of deleting inhabited land are going to be far more accurate and useful than maps that incorporate historical sites, habitat associations in states of disrepair, and intervening territory between inhabited sites, because the first type of error will be trivial compared to the second.

These statements are not meant to excuse error or imply that it is acceptable. My goal, as with anyone attempting to document such information, is to present absolutely accurate maps. I hope that knowledgeable readers will help identify areas I have overlooked for some species. Certainly one of the most useful aspects of a distribution map is to spur closer examination of areas both included and eliminated as part of a species' range; thus such maps serve as a flawed framework onto which another student of the organism can hone a truer vision of its geographic distribution.

1

What Are the Southern Pine Woods?

The biological communities here termed "southern pine woods" are contained within a clearly definable physiographic region and host a unique assemblage of flora and fauna. The southern pine woods are those natural areas visually dominated by an overstory of pine and occurring along the Atlantic and Gulf coastal plains, including peninsular Florida. This habitat, in all its variations, was the dominant ecosystem throughout the southeastern United States until the advent of the lumber rush of the early twentieth century. The surviving remnants of these great forests are scattered from North Carolina southward through coastal South Carolina, southeastern Georgia, and much of Florida including the Keys. Increasingly more scattered examples of pine woods are seen westward along the Alabama and Mississippi Gulf coast, through central Louisiana, finally reaching their natural limit at the eastern edge of the prairie in east Texas. The original range of the southern pine woods is still implied by present-day fragments, but it blanketed the land in vast, largely unbroken swaths quite unlike the splintered pieces that remain.

Subdivisions and Types

Biologists try to organize landscapes by defining categories of habitat and arrange them as vegetational assemblages. Defining boundary lines between the zones according to the dominant vegetation is a useful approach, because the flora reflects influential environmental characteristics such as substrate moisture, soil type, and climate that help determine what animal species are able to flourish. However, no single scheme best defines ecological assemblages for all types of animals. For example, ground-dwelling amphibians are dependent on the presence of standing water, at least seasonally, so factors affecting soil drainage are especially important to them. For pine

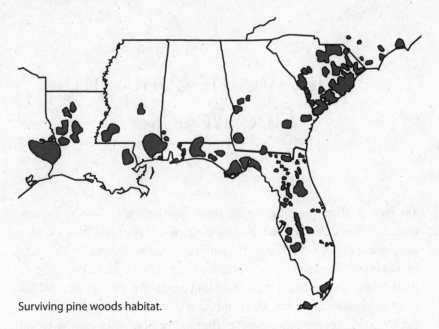

Surviving pine woods habitat.

woods amphibians, then, the criteria for subdividing habitat types would need to emphasize factors such as substrate material and slope, things that would be marginally relevant to an ornithologist wishing to make sense of the distribution of birds across a geographical area. Although a change in forest structure from deciduous to conifer-dominated community is obvious even to a casual observer, appreciating subtle differences between the several southern pine woods habitats requires close examination.

To gain a clear insight into the distributional patterns of reptiles and amphibians within the pine woods and discern the subtleties that differentiate the habitats from the perspective of their reptile and amphibian inhabitants, several factors need to be considered. First, the abundance and dependability of moisture is critical for pine woods herpetofauna. Ground moisture variability at the local scale is determined chiefly by two factors: substrate material and slope. Pine woods substrates tend to be porous and quick draining. Consequently, most habitats are dry and parched much of the year. Anywhere the terrain is hilly, pine woods habitats are xeric. At times, many pine woods ridges experience desertlike conditions, and this quality is revealed in patches of prickly pear cactus and crusty lichens clutching sun-baked sand (plate 1.1).

The amount of sun reaching the forest floor is a critical determinant of the particular species of reptiles and amphibians inhabiting an area.

Preferred body temperatures of amphibians are much lower than those of reptiles, and amphibians are much more prone to desiccation; therefore, they prefer the more shaded and mesic types of pine woods. The most important environmental factor determining the amount of sun exposure on the ground is canopy coverage. Areas where the pine overstory is closed or nearly so are faunistically and ecologically different from more sparsely shaded pine woods, and are here termed "forest"—using this term in a narrow sense—to indicate this vegetative condition. A dense canopy helps to maintain cooler, moister substrate environments conducive to amphibians and some types of reptiles (plate 1.2).

As a result of the very low fertility of some pine woods soils, which support only sparse coverage of dwarfed pines, coupled with the frequent fires that serve to sweep away undergrowth, many areas look patchy and rather scruffy, quite distinct from forest, and are termed scrub. As with pine forest, slight variations of scrub pine communities can be found within the larger vegetative community types outlined below. Thus, one may divide pine woods habitats to delineate distinctive herpetofaunal zones into flatwood forest or scrub, savanna forest or scrub, ridge forest or scrub, and so on. Scrub situations are both the more common, and the most distinctive, of the two categories in the four natural biozones described below. Only under human management constraints, in pine plantations, have the southern pine woods been obliged to persist as forest (plate 1.3).

Substrate material, slope, and overstory coverage: by organizing pine woods according to the qualities and proportions of these environmental variables, a simple but informative arrangement of reptile and amphibian habitat types in the pine woods comes into view.

Pine Flatwoods

Flatwoods present the most dynamic hydrology of any of the major habitat divisions. When rainfall is high, flatwoods take on a swampy character and provide excellent sites for amphibian reproduction, since they have been dry for the preceding months and contain no predatory fish. During regular drier periods, flatwoods become as parched as any other pine habitat. The generous but transitory availability of abundant standing water results in dramatic shifts in amphibian and, to a lesser extent, reptile activity. During certain times of the year, flatwoods have a more conspicuous amphibian presence than any other pine woods association, but at other times, frogs, toads, and salamanders are as secretive as their counterparts in more consistently dry habitat. Wintertime is an especially good time to survey for flatwood amphibians, as most breeding takes place then and adults are readily found in accumulations at breeding ponds (plate 1.4).

Flatwoods develop where drainage is poor. In keeping with their name, flatwoods are very level; slope is either nonexistent or conducive to collection of runoff in slight basins that form over certain underlying soils or base rock structures. The habitat is promoted when the overlying surface soils, which may be quite sandy and porous, are on a foundation of hard-packed clay, which hampers percolation. These basic parameters are found mainly in pine woods along the coastal plain; thus herpetofauna with more inland centers of distribution such as *Plethodon kisatchie*, *Stilosoma extenuatum*, and *Pituophis ruthveni* are rarely if ever found in flatwoods. One would expect that the species most dependent on the consistent presence of standing water would be most closely tied to flatwoods associations, and this assumption is confirmed by the presence of one of the most aquatic endemics, the Striped Newt, *Notophthalmus perstriatus* (plate 1.5).

Superficially, some flatwoods resemble savanna but differ in being more consistently mesic. In both communities, the lack of landscape relief creates open vistas through the open midstory zone. A key characteristic that helps to differentiate these two divisions is the presence of saw palmetto, *Serena repens*, which is nearly ubiquitous as the dominant understory vegetation of flatwoods. In the drier savanna associations, grasses fill this understory niche (plate 1.6).

Pine Savanna

Designating dry, flat pinelands as savanna imparts an accurate image of what most of these sites look like. Most readers are familiar with the term "African savanna," conjuring images of vast, sweeping vistas of tall grass and only a sparse scattering of trees and occasional clumps of bushes, creating a parched and sun baked, but not desolate environment. The pine savannas of the southeastern United States are, or more accurately once were, similar in appearance, if less grand in scale (plate 1.7).

Savannas and flatwoods are distinct environments in terms of habitat offerings for reptiles and amphibians, but the cause of this contrast is simple and basic—drainage. Savannas are often at greater elevation than flatwoods, and their topography differs in that rainfall runoff is facilitated and they are not situated in locations where flooding is likely to occur. Consequently, savanna vegetation is less moisture tolerant than the flora typical of low-lying flatwoods. The hallmark of pine savanna is a carpet of grass dominating the ground vegetation, but these plants do not tolerate flooding well, so in mesic pinelands they are replaced by other flora types. Throughout much of the southeast, wiregrass (*Aristida beyrichiana*) is the species that blankets great swaths of savanna ground. West of the Mississippi River, the pine savannas of Louisiana and east Texas support a ground cover of bluestem grass (*Andropogon girardi*). Both savanna forest and scrub variations frequently develop (plate 1.8).

Being a somewhat broader designation than the two habitat divisions at opposite ends of the moisture gradient—ridges and flatwoods—savanna is less rigidly definable and thus the more common and widespread type of pine woods. In other words, if a pine-dominated landscape is not so low-lying as to become periodically swampy during periods of heavy rain, and not so high and sloping that only a sparse and xerophilic ground cover exists, then herpetofaunal components characteristic of savanna will occur.

The most impressive expanse of good pine savanna habitat is spread across three adjacent tracts running from southern Alabama across the Florida panhandle to the coast. This natural treasure exists as a result of combined management practices applied in Conecuh National Forest, Blackwater River State Forest, and Eglin Air Force Base. In fact, this combined area of protected habitat is the largest stand of longleaf pine savanna in existence, encompassing more than 283,000 ha (Compton 2000) of pineland refuge.

The composition of the characteristic herpetofauna reflects the intermediate position of savannas on the annual moisture gradient. Free of both frequent parching and saturation, savannas offer a milder environment for reptiles and amphibians. Water is neither a scarce nor overwhelmingly predominant feature, so while many amphibians inhabit savannas, the more delicate and moisture-dependent species, newts and Chorus Frogs for example, are more abundant in flatwoods. Reptiles adapted to xeric pine uplands, such as Eastern Pine Snakes and Florida Worm Lizards, will be found in savannas on occasion, but are more dependably encountered along drier barren ridges.

Pine Ridges

Whereas pine woods amphibian fauna reaches its greatest diversity in flatwood situations, pine ridges harbor the largest endemic reptilian component. Ridges arise from a variety of mechanisms, but for the purpose of understanding the herpetofauna, two categories are most relevant. Some ridges are composed of deeper and sandier substrate than others. Particularly those lying in close proximity to the present-day coastline, these topographic features are relic dunes from prehistoric times, when sea levels were much higher than today. As the oceans receded, they left behind these products of wave and wind action. Ridges produced in this way reveal themselves by being directionally aligned in parallel orientation over a broad geographic region. For example, the dry ridges in north central Louisiana run in an easterly or northeasterly direction overall; those in central Florida are grossly aligned north to south, each thus reflecting the former presence of a coastal terrace of beach and dunes. The reason this phenomenon is instructive to the student of pine woods zoogeography is that some taxa evolved in these ancient coastal areas under distinctive selective pressures. Although the expanding mainland has long since subsumed the sites, the endemic herpetofauna still reflect this singular pedigree in their unique and divergent characters (plate 1.9).

Other ridges supporting pine woods, mainly those well inland, cannot be explained by the former presence of the sea. Ridges such as these are often little more than that—longitudinal elevations where the physical and biotic components vary only slightly from those in the general surroundings (plate 1.10).

A variety of ridge habitat associated with maritime environments is distinctive enough, from a herpetofaunal perspective, to warrant special mention. Behind beachfront dunes, wherever elevations are sufficient to keep saltwater tables well below the surface, scrubby stands of sand, slash, or rarely longleaf pine may be found, along with a small assemblage of reptile and amphibian species. The terrain is usually high and rolling, because this habitat has evolved on top of windswept dunes. The loose sand along the numerous slopes is virtually devoid of any vegetation; coupled with the continual sea breeze, it yields an environment that is harshly xeric and constraining to reptile and amphibian life. Because of the close proximity of the sea and the exceptionally deep sands, these areas are usually open and sunbaked scrub, with standing freshwater only intermittently available. None of the species described in this book are specialists for this type of habitat, but several may be found here on occasion, though less abundantly than in their more typical associations. In fact, any of the ridge specialists whose distribution meets the coastline can be expected in coastal strand scrub.

Among the small cadre of amphibians hardy enough to exploit these spaces, the most likely to be encountered is the Oak Toad, *Anaxyrus quer-*

cicus, and the Pine Woods Treefrog, *Hyla femoralis*, both of which may be found within yards of the tide line. Ridge specialist reptiles found along the coast include the Southern Hognose Snake, *Heterodon simus*, and both sub-species of Eastern Pine Snake, *Pituophis melanoleucus*. One other species, paradoxically more typical of mesic flatwoods situations, is also abundant in dry coastal scrub, and that is the Gulf Coast Box Turtle, *Terrapene carolina major*, which thrives on coastal dunes along the Apalachicola coast of Florida. (plate 1.11).

Few pristine tracts large enough to function as ecosystems remain within their historic span across the southeastern coastline, from North Carolina to Mississippi east of the Mississippi River delta. From a human perspective, coastal strand scrub is synonymous with "beachfront property," and the few areas not completely developed are imparted with desirable designations such as the next "undiscovered" coastline or the last remnant of unspoiled vacation paradise. In a few cases, this designation has had the beneficial effect of spurring habitat acquisitions by conservation groups, such as the Bon Secour National Wildlife Refuge parcels that now protect virtually all the remaining coastal pine woods in Alabama (plate 1.12).

Pine Rockland

The pine woods ecosystem with the most limited distribution is the pecu-liar pine scrub in southernmost Florida. This unique association of slash pine scrub atop shallow oolitic limestone bedrock is restricted to southern Dade County, Key Largo, and some of the lower keys, including Big Pine, No Name, Little Torch, Middle Torch, Ramrod, Cudjoe, Summerland, and Sugarloaf. Originally occupying more 81,000 ha, pine rockland has been reduced to shattered fragments owing to the tremendous value of this real estate for residential and vacation industry development. The ecosystem persists in four general areas today: a patchwork of approximately 15 public and privately owned parks and preserves in Dade County, a large contiguous tract in Everglades National Park, a small portion of Key Largo, and within Key Deer National Wildlife Refuge on Big Pine Key. Many of these parcels are too small to function as ecosystems, and merely provide a visual representation of what such habitat once looked like. Others, such as the larger tracts around Miami's Metrozoo and on Key Deer National Wildlife Refuge, continue to provide refuge for the full array of pine rockland flora and fauna, but do so without connecting corridors and in isolation, rendering them vulnerable to sudden environmental stress (plate 1.13).

Pine rockland is distinguished by its substrate of exposed limestone. Unlike terrestrial microhabitats composed of sand and decaying leaf litter, pine rockland offers fewer mesic refugia for small animals. The xeric, sun-baked ground is covered with an abundance of limestone rocks and boulders, or composed entirely of monolithic bedrock cracked and fissured and providing miserly escape from the sun for reptiles and amphibians. In the few areas where soil lies, it is gravelly, shallow, and hard packed. There are few fallen trees, and those that do exist are often dry to their cores. The dead leaves of Thatch and Silver Palms (*Coccothrinax argentata*) afford hiding places unique to this habitat (plate 1.14).

Particularly in the Florida Keys, rockland lies in close proximity to coastlines and estuarine habitat, and freshwater is scarce. Reptiles satisfy their water needs primarily from the frequent rainfall that characterizes the subtropical climate. Not surprisingly, amphibians are poorly represented in this harsh environment, particularly those requiring reliable aquatic nurseries for egg and larval development. The most prevalent source of standing water are the shallow depressions called solution holes, which are basins resulting from the dissolution of underlying stone. These catchments, often less than two meters across and only a few centimeters deep, collect rainwater and provide ephemeral pools (plate 1.15).

Although open scrub conditions are maintained typically by periodic fire sweeping across the infertile ground, pine rockland is the only pine habitat routinely cleared by the sea. Low-lying tracts, particularly in the Keys, are subjected to overwash during tropical storms. Once inundated by salt water, some of the vegetation dies back, including the slash pine overstory. Some sites on Big Pine Key are so sparsely treed that they resemble cut-over pine plantations (plate 1.16).

Pine rockland is the only pine woods variant that in its entirety is both interspersed with and completely surrounded by large-scale human activity. It is a wild land but completely landlocked by development. All the herpetofauna living in this area do so with the high likelihood of encountering humans and living alongside their dwellings. They have exploited this situation in various ways, finding artificial refuge in trash dumps or under backyard debris. The wildlife utilizes intermittent manmade sources of water such as old trenches dug for mosquito control schemes or rainwater receptacles like old tires. In spite of this, the pine rockland endemics are the most imperiled of all pine woods herpetofauna (plate 1.17).

Pine Plantation

Vast areas of the rural gulf coastal plain are cultivated as commercial tree farms. Although all have the goal of producing a steady flow of softwood timber for sawmills, the management protocols in force vary greatly, making generalizations impossible regarding the compatibility of this type of habitat with herpetofaunal biodiversity. At one extreme, plantations may be no more than loblolly tree farms, huge expanses of aggressively managed monoculture. All other flora is vigorously suppressed to reduce competition, fire is eliminated as a risk factor to young trees, and large tracts undergo brutal clear-cutting regimes that are so frequent as to keep the entire area under constant disturbance. This form of silviculture maintains a harsh environment for most wildlife. Particularly vulnerable are the relatively sedentary and habitat-sensitive reptiles and amphibians. Although such areas may appear to be forests to uninformed motorists driving through the miles of loblolly-lined roads that transect these operations, they are in reality biological wastelands and disgraceful displays of mankind's disregard for the natural environment (plate 1.18).

Fortunately, more often than is appreciated, many commercial silviculture tracts do not stray far from the natural state, and do support a full component of herpetofauna. Despite concern over the cost and liability of prescribed burning, some timber companies are including it in their management plans, simultaneously reducing the use of herbicides that suppress undergrowth. Periodic thinning, if done carefully so as not to demolish the structure of the understory vegetation and surface features, helps to maintain the open canopy necessary for a sunny forest floor on which many pine woods reptiles depend. Clear-cutting is either not employed at all, with trees harvested on a selected basis, as is often the case when mature saw timber is the desired end product, or practiced on a limited scale with an eye to timing and spacing so that bare patches of ground nowhere dominate the landscape. Plantations, when managed with environmental sensitivity, can be simultaneously profitable and quite conducive to the maintenance of herpetofauna, even improving habitat for some taxa by amplifying ecotones and offering a more varied mosaic of stand age and compositions than is present in unmanaged sites (plate 1.19).

Southern pine plantations run the gamut from poor replicas of forest, virtually devoid of reptiles and amphibians and well-deserving of the accurately ugly epithet "industrial forest," to conscientiously tended pine woods that effectively serve the dual purposes of lumber resource and wildlife reserve. However, because of the wide range of priorities and practices that govern these lands, one can find plantations that offer some of the best opportunities for observing endemic herpetofauna in close proximity to some of the most discouraging and depressing examples of natural heritage mismanagement to be found anywhere. So as to temper too much optimism, it should always be remembered that even the most responsibly managed pine plantations are being maintained for one overriding purpose—the profitable production of forest products. Therefore, these areas are highly unstable in nature; economic upturns and downfalls, changing market dynamics, waxing and waning of demand for a particular product, corporate reorganizations, land sales between companies, and the inevitable changes in key personnel: each of these has the capacity to completely transform land management philosophy and priorities governing a particular plot virtually overnight. Thus it will always be preferable to ensure that key tracts representing flatwood, savanna, ridge, and rockland habitats as well as critical habitat for rare species are generously sequestered into public stewardship as buffers against trends and fashions in conservation that will come and go. Still, it must be said that commercial silviculture and wildlife conservation are not incompatible endeavors, and I know of no more heartening realization when I ponder the dire state in which so many pineland communities find themselves today.

Regional Endemism

Another way to organize the southern pine woods from a herpetological perspective is to define biogeographical zones. These are regional delineations that encompass a recognizable herpetofaunal component, one in which unique species occur. These areas may include several of the habitat subdivisions previously described, but are nonetheless cohesive biological units that stand apart from the remainder of the southern pinelands by virtue of the unique reptile and amphibian forms which have evolved there. Within the broad parameters of the southern pine woods generally, four regions are the result of unique physiographic histories and consequently contain a reptile and amphibian fauna found nowhere else on earth.

Miami Rock Ridge and the Lower Keys

Geographically isolated from other sites, the rocklands of south Florida and the Keys are also the most biologically distinct. Pine rockland originally dominated the natural environment of extreme southern Florida, including portions of present-day Broward County and most of Dade County, terminating at Long Pine Key. The habitat gave way to the grassy wetlands and hammocks of the Everglades, resuming again at Key Largo and continuing sporadically through to the Lower Keys. Today, only remnants of this habitat remain in Broward and Dade counties, and the Key Largo pine rocklands are gone. Miami rim rocklands and the pine rocklands of the Lower Keys have a distinctly tropical character owing to the abundance of various species of palms in the understory vegetation. These environments are also noteworthy for the abundance of endemic plant species they support.

Somewhat paradoxically, the number of endemic reptile and amphibian taxa is lower than in the other biozones. One reason for this may be the insular genesis of the land, accentuated by the harshness of the environment. Among amphibians endemic to pine wood areas, only the Oak Toad ventures into these dry scrubs. Only two endemic reptiles occur here, the Key Ringneck Snake and the Rim Rock Crowned Snake, but because both are rare, the south Florida rocklands are very significant concerns for conservation biologists (plate 1.20).

Central Florida Ridges

A series of eroded sand ridges in central Florida are the faintly perceptible remnants of a prehistoric coastline, but the bioregion that has evolved on these features is spectacular. The physiography of the central Florida ridge country was shaped by changing levels of land and sea such that islands sequentially formed, variously joined and separated, and finally coalesced into present-day unity with the continental land mass. Beginning in the Miocene, some of the terrain of what is now inland central Florida was the coastline of an ancient sea. Wind and wave action built high sand dunes along an approximately 160 km stretch of prehistoric beachfront. Some of the fauna and flora now residing exclusively along this strip of habitat are the descendants of the unique forms that evolved there. Although fluctuating sea levels alternately transformed these lands into isolated islands in a shallow ocean, only to be reconnected later, the ridges themselves persisted throughout long periods of geological time when virtually all the remaining lands that surround them today were periodically submerged. Thus these ridges, and the plants and animals that distinguish them, represent an old lineage compared with other species inhabiting present-day Florida.

Running north to south through the middle of the Florida peninsula, five major ridges bear names: Lakeland, Winter Haven, Bombing Range, Lake Henry, and the largest and most diverse in terms of habitats, the Lake Wales Ridge. Several rare habitats are concentrated along these high sand hills. Much of the area is given over to open scrub with little or no large tree cover that supports its own unique herpetofaunal elements distinct from those of the pine-dominated habitats. The Sand Skink, *Neoseps reynoldsi*, and the Florida Scrub Lizard, *Sceloporus woodi,* favor the open, desertlike climate and deep, loose sand substrates of the Rosemary Scrubs, although both may be found in pine woods situations as well. Atop the crest of the ridges is where the largest stands of sand and longleaf pine wood relicts can still be found, and are the preferred habitat of two unique Florida ridge inhabitants. The Short-tailed Snake, *Stilosoma extenuatum*, and the Florida Worm Lizard, *Rhineura floridana*, are singular presences on the North American herpetofaunal list at the generic and familial level, respectively, reflecting their long isolation and independent evolution. Other distinctive subspecies have evolved on these ridges, such as the Mole Skink, *Eumeces egregius*, and the Florida Crowned Snake, *Tantilla relicta*.

The central Florida ridge country is under tremendous pressure by the burgeoning human population. The picturesque landscape is ideally located for real estate development and this is occurring at a rapid rate. The dry and fast-draining soils spare developers many of the headaches associated with construction in other parts of Florida, where perpetually soggy ground, periodic flooding, and the potential of direct collision with tropical storms conspire to complicate building projects and increase costs. These lands are also at the epicenter of the Florida citrus industry, and the majority of the acreage of this biological zone has been converted to this purpose.

Apalachicola Flatwoods

In similar fashion to the mechanism that created the endemic herpetofauna of Florida's central sand ridges, the flatwoods in and adjacent to Apalachicola National Forest have generated their own distinctive forms. During roughly the same period of high sea levels that inundated peninsular Florida and fragmented high terrain into isolated islands, the area now comprising the Apalachicola River basin was largely submerged (Brenneman and Tanner 1958). Only a few land features visible today may have been continuously above sea level since the Pleistocene. It is postulated that several sand ridges lying northeast of the town of Apalachicola are the remnants of ancient barrier islands upon which the evolution of some of the Apalachicola region's unique plants and animals took place (Means and Krysko 2001) (plate 1.21).

Although Florida's central ridges and Apalachicola's flatwoods share a common genesis, their subsequent development followed distinct pathways, resulting in contrasting habitats and distinctive herpetofaunal elements. While the central ridges became deeply landlocked and thus impervious to coastal erosive forces, the Apalachicola region remained vulnerable to such action. Even after sea levels fell and the barrier islands congealed into the mainland approximately 5,000–6,000 years ago, the landscape continued to be lowered by the dynamics of rainfall and ocean flooding. Unlike the xeric central ridges, the Apalachicola flatwoods today are traversed by several large, meandering rivers and a multitude of small creeks and sloughs. The rivers have changed course during the past several thousand years (Donoghue 1989), further accentuating erosion of high ground and the deposition of alluvial top soils. Today, the Apalachicola flatwoods retain a soggy character with a diverse matrix of intertwined wetlands and seasonally flooded pine flatwoods. The region harbors numerous boggy situations and permanently aquatic environments in close proximity to tracts as dry as any found along the Lake Wales Ridge, quite unlike the overwhelmingly desertlike conditions that prevail in central Florida (plate 1.22).

The unique combination of a past existence as xeric coastal islands followed by a transformation into coastal mainland wetlands has produced singular consequences in the herpetofauna of Apalachicola. In contrast to the oddities of the dry ridges, which are primarily adapted to fossorial lifestyles within the loose sands, the endemic herpetofauna of Apalachicola are larger, less secretive, and more diurnal. None, however, has evolved to species-level distinction, perhaps reflecting less impervious geophysical isolating barriers to genetic interflow with neighboring taxa. Nowhere else in the southern pine woods can one encounter the Brownchin Racer, *Coluber constrictor helvigularis*, the Apalachicola Lowland Kingsnake, *Lampropeltis getula meansi*, or the Gulf Coast Box Turtle, *Terrapene carolina major*, in its purest and most divergent form.

Trans-Mississippi Sandhills

The central Florida ridges and Apalachicola flatwoods are remarkable for their genesis through paleohistoric isolation from terrain they presently abut. A similarly formed expanse of pine woods lying west of the Mississippi River is even more remarkable in having secondarily returned to geographical isolation from other pine woods habitat, prompting a new era of independent evolutionary trajectories of its herpetofauna. The area called the trans-Mississippi sandhills is situated between the western outpost of the pine woods and the east Texas prairie lands. At its southern border, near Vernon and Rapides parishes in Louisiana and Orange County, Texas, a rapid transition to low lying wetlands and estuarine habitat contains it. To the north are the rolling deciduous woodlands of the Ouachita Mountain foothills, with one last relict habitat resembling southern pine forest contained in the shortleaf pine hill forests in Columbia, Union, Bradley, and Cleveland counties in Arkansas. The trans-Mississippi sandhills have evolved in isolation from the maternal forest to the east as a result of intervening development of the Mississippi River and associated bottomland riverine forest habitats (plate 1.23).

The Mississippi River has been a barrier to genetic interflow between populations of reptiles and amphibians for approximately 12,000 years. The extent to which the faunas to the east and west of the Mississippi have come to differ in such a seemingly short time is illustrated with a comparison of the pine woods herpetofaunas. In southeastern Louisiana, the so-called Florida Parishes, lying east of the Mississippi and north of lakes Maurepas and Pontchartrain, have in recent times harbored species apparently always lacking in habitats west of the River, including the Black Pine Snake, Eastern Diamondback Rattlesnake, Gopher Tortoise, and Dusky Gopher Frog. The trans-Mississippi sandhills, on the other hand, have hosted the evolution of species not found further east, including the Louisiana Pine Snake, Slowinski's Corn Snake, and Louisiana Slimy Salamander. Although fully as distinctive as the other three geographical concentrations of herpetofaunal endemism, the trans-Mississippi sandhills are the least appreciated and valued, and their herpetofauna receives the weakest level of protection at both the state and federal levels.

2

Observing Reptiles
and Amphibians

I encourage you to seek out the pine woods and explore them in search of the reptiles and amphibians described in these pages. The average person rarely sees most of these creatures. Only people made aware of their existence and educated in their habits and habitat preferences are likely to notice the majority of pine woods reptiles and amphibians. This herpetofauna must be laboriously sought to be found, and in this chapter, I offer tips for how best to encounter them.

I implore all those who take my suggestion to go looking for herpetofauna to do only that, and not collect them. These are fragile lands, and most pine woods habitats that still harbor their endemics are small, isolated relicts of a formerly glorious past now long gone. Most of the reptile and amphibian populations are small, and removing even a few could result in greater vulnerability of the remaining animals to the multitude of other pressures increasingly bearing down on them. Consider yourself very fortunate if you ever see a Mimic Glass Lizard, a Tan Racer, or a Striped Newt in the wild. I sincerely encourage you to seek out these experiences whenever you can, but please, leave the forest exactly as you found it, including the individual organisms that live there.

Even those who heed the ethic of leaving herpetofauna as found in situ could do great damage to habitat without intending to. The most fragile part of any pine woods is the uppermost layer of the ground. This thin crust is home to many of the creatures highlighted in these pages, and in more intricate ways than simply as the substrate that they scurry, hop, or crawl across. The first several centimeters of forest floor, including low-lying ground vegetation, the layer of fallen pine needles, and the loose sand they cover, is where small fossorial snakes such as Crowned Snakes, Ringneck Snakes, and Short-tailed Snakes prowl. This delicate layer is where the Striped Newt and Flatwoods Salamander find mesic refuges during the precarious time spent in dry uplands between breeding events. The leaf litter and loose sand are where Mole Skinks find nesting sites in an inhospitable landscape and where Florida Worm Lizards find their wolf spider prey, arthropods that are themselves more inextricably tied to the unique substrate characteristics of their habitat than the lizard itself. Hiking carelessly over this land can wreak havoc on microenvironments with deceptively little sign. Pay close attention to the crunching sounds of lichens being destroyed, or the sensation of feet sinking into the sand as they collapse animal burrows, or the sight of oak leaves that had captured the morning dew now irreparably ground into the sand of a xeric ridge slope, and you can see the damage that one well-intentioned hiker can do. Please stay on established trails when exploring the southern pine woods. By doing so, you not only avoid causing damage yourself, but also keep footpaths clear and conspicuous, thus encouraging those who follow to do the same.

Throughout this book I describe the specific microhabitats and preferred sheltering places of reptiles and amphibians. Many species are fond of secreting themselves under objects such as fallen trees, dead palm fronds, or jumbles of limestone rock. Often the only way to find a particular species is to spend hours turning over every such object that can be found. This practice can be done either with respect for the environment, or in a brutish and inconsiderate way. The proper way to look for organisms under objects is by very carefully replacing the object in as close to its original position as possible. If done correctly, it should be impossible to tell that the object has been moved. Too many times I have walked along a foot trail and seen logs or rocks left upturned after some thoughtless person has preceded me. Such people may be reptile enthusiasts, but they are not naturalists in any sense. When a reptile or amphibian is discovered under an object, it should be gently removed before the cover is set back in place, then placed at the edge of the object so it can crawl back beneath on its own, to avoid being pinned or trapped.

Hacking apart rotting logs should be avoided, though this will make finding some species more difficult. Fallen trees that hold moisture are important features to many endemics such as the Louisiana Slimy Salamander, Mabee's Salamander, and the Pine Woods Snake. It takes a long time for a fallen trunk to "age" properly before it becomes a haven for small herpetofauna, and an ignorant individual can destroy all such refuges over several acres in very little time.

One way an observer of reptiles and amphibians can engage in searching for secretive species without damaging the environment is to put out "cover boards." Items such as scrap plywood, plastic siding, or roofing tin can be salvaged from dumps and spread across the ground in areas likely to harbor rare species. After several months or longer, the cover can be lifted and checked for animals that have sought refuge beneath. In this way a person is doing no damage to ground cover and is perhaps even improving habitat by increasing the amount of refugia in the vicinity.

Locating Suitable Habitat

As used here, "southern pine woods" refers to a biological community in the United States where the most visible biotic feature is a native species of pine tree. True pine woods, where the species highlighted in this book are found, maintain their unique appearance and characteristics through natural dynamic processes or human-controlled regimens designed to mimic these processes. Most of the vast swaths of land seen throughout the South that are dominated by pine trees—those that line the highways along the more rural stretches, for example—are not native pine woods communities but merely concentrations of pine trees. They are the result of someone having planted pine where none formerly existed, or through cutting, poisoning, or otherwise eliminating all other floral elements in natural pine woods until a pine monoculture has been created. Other piney areas may once have been natural sites, but through fire exclusion, alteration of drainage patterns, or the careless placement of roads that isolate tracts into nonfunctioning ecosystems, they are no longer pine woods. Searching for endemic reptiles and amphibians in these places is futile.

Southern pine woods are very attractive to developers. They are high and dry, resistant to flooding even when situated along the coast, and often flat and easy to clear. Being such a scarce and valuable commodity, a natural pine ecosystem in the southern United States never exists by accident. These biotic communities persist only when their stewards value their existence and specifically manage them for preservation. In only three types of situations are such policies applied on a large enough scale to foster viable ecosystems: publicly owned natural areas, private preserves, and, surprisingly, military bases and training grounds.

Public Lands

Most natural pine woods today are located within the National Forest system. This vast collection of undeveloped lands is truly the salvation of the southern pine woods, because without it there would remain only a sparse scattering of examples, separated by great distances. Our National Forests are also an essential part of the conservation of endemic pine woods reptiles and amphibians, because no species is without significant populations located in them. In Louisiana, Slowinski's Corn Snake and the Louisiana Slimy Salamander have the great majority of their distributions encompassed by Kisatchie National Forest. The majority of specimens of Apalachicola Lowland Kingsnakes have been taken in Apalachicola National Forest, with most of the balance coming from other public natural areas. One species, the Dusky Gopher Frog, is now known only on National Forest land, in the southern sections of De Soto National Forest in Mississippi. Other National Forests that contain important areas of pine woods are Croatan in North Carolina, Francis Marion in South Carolina, Osceola in Florida, Conecuh in Alabama, Homochitto in Mississippi, and Sabine and Angelina National Forests in Texas.

Other types of public lands play an important role in pine woods herpetofauna conservation. National Wildlife Refuges protect significant tracts. The 5,000 ha St. Vincent National Wildlife Refuge is a gem of natural coastal strand, flatwoods, and savanna habitats, and contains a healthy population of Gulf Coast Box Turtles in their purest and most spectacular form, as well as Brownchin Racers and Oak Toads (Irwin et al. 2001). Bon Secour National Wildlife Refuge in Baldwin County, Alabama preserves the last functioning ecosystem of coastal pine flatwoods remaining on the state's coastline, most of which would have been destroyed by the real estate boom that is taking place there. Oak Toads, Pine Woods Treefrogs, and perhaps rare species such as the Mimic Glass Lizard and Flatwoods Salamander still make their home in Bon Secour, their last stand in the region as they are surrounded by condominiums and beach homes. Perhaps the most impressive example of a National Wildlife Refuge's contribution to pine woods herpetofauna conservation is Key Deer National Wildlife Refuge in Big Pine Key, Florida. On this island, the majority of pine rockland is preserved and skillfully managed by forest stewards, and here lives the only known sustaining population of the Key Ringneck Snake. Other important National Wildlife Refuges for pineland herpetofauna are the Big Thicket Preserve in Texas and Carolina Sandhills National Wildlife Refuge in North Carolina. Many National Wildlife Refuges are rather small and often juxtapose developed areas, which imparts both great value and daunting challenges to wildlife conservation (plate 2.1).

State parks and forests, usually smaller than National Forests, play an important role by providing corridors between other protected tracts and by protecting especially critical small sites. Blackwater River State Forest in Santa Rosa and Okaloosa counties, Florida, plays a major role in conserving pine ecosystems by connecting the vast holdings of Eglin Air Force Base and Conecuh National Forest. These three areas combined represent the largest expanse of longleaf pine ecocommunity that remains. The State of Florida has an extensive system of state parks and forests, and many have intact piney habitats. Particularly large and fine examples include Pine Log, Tate's Hell, and Myakka River. Other important state-managed lands are Bladen Lakes State Forest in North Carolina, Sandhills State Forest in South Carolina, Waycross and General Coffee State Forests in Georgia, and Gulf State Park in Alabama.

Military Reservations

The most surprising areas where pine woods herpetofauna find refuge are military grounds. A serendipitous but fortunate set of circumstances has made military bases and their surrounding lands an integral part of the conservation of these organisms. The training of thousands of people in the techniques of warfare, and the testing of weapons, require very large areas, and bases such as Fort Polk, Eglin Air Force Base, and Fort Stewart rival National Forest districts in the acreage they encompass. Unlike state or private reserves, which are often rather small, some military bases encompassing pine habitat are vast enough to preserve whole functioning ecosystems, from the level of apex predator through inclusion of a full complement of invertebrate and floral diversity. Because troops are being trained under conditions similar to those they will experience in real conflicts around the world, forests are valued as assets and have been left largely unmolested. The bombing ranges at these sites also promote healthy pine ecosystems. The small-scale fires often caused by detonations have kept portions of these lands under a more consistent burning regimen than many other types of federal lands originally set aside for the primary purpose of natural area preservation. The latter have usually been maintained under protocols that prohibited fire during some management tenures. It is actually easier to find virgin stands of native pine on some military reservations than it is on National Forest land. Finally, although limited areas of most military land are made accessible for public recreation, the control and strict oversight of these activities and the fact that large sections are off-limits to hikers, dirt-bikers, reptile collectors, and the like keeps degrading activities to a minimum. Misuse by visitors, such as off-road vehicle damage, trespassing the boundaries of sensitive sites like amphibian breeding pools, or poaching, is much easier to control on military-managed pine woods than in National Forests.

As a result of this unique set of characteristics, the most expansive, pristine, and biodiverse pine woods in the South are under the stewardship of the U.S. Department of Defense. Excellent examples of habitat typical of their region are found in most southern states. In Louisiana's Fort Polk, Slimy Salamanders and Slowinski's Corn Snakes exist in abundance. Here also the Louisiana Pine Snake, America's rarest species of large snake, makes one of its last stands in the wild. Camp Shelby, in Mississippi, is home to a vigorous and probably the largest extant population of Black Pine Snakes. Camp Shelby has also been the site of important research into the ecology of this taxon. Surely the crown jewel of all southern pine preserves, under any management structure, is Eglin Air Force Base in Okaloosa and Walton counties, Florida. At Eglin, not only are magnificent examples of ridge, savanna, and flatwood habitats available to be enjoyed, but the level of stewardship of these ecosystems is maintained at a level of integrity that sets the standard. In Georgia, two bases at opposite sides of the state contain excellent tracts of habitat. Fort Benning, in west Georgia, preserves many upland sites, while Fort Stewart, in the southeastern quadrant, contains both xeric and flatwood habitat, including an important metapopulation of Striped Newts.

Although not a typical military facility, the Savannah River Site (SRS) supports many pine woods ecosystems on its closely monitored lands, and plays a pivotal role in herpetological research generally, and pine woods herpetology particularly, by hosting the University of Georgia's Savannah River Ecology Lab (SREL). The SREL mission is to study the ecological effects of the SRS operations. Over the 50-plus years of its existence, the SREL has yielded a massive body of herpetological research and a small army of distinguished herpetologists.

Private Preserves

A diverse array of privately owned pine woods scattered throughout the southeastern United States provides additional refuge for herpetofauna. Often these are small parcels not being utilized by the owner and simply able to perpetuate in a natural state. Such scenarios are tenuous, and I have several times observed the loss of these habitats as they fall victim to the capricious nature of private land ownership and the changing priorities of the owners.

Microhabitats

Even if good habitat is located and entered, the secretive nature and micro-habitat specificity of most pine ecosystem reptiles and amphibians make it difficult to find specimens without an educated search. Key to finding each of the species presented in this book is an understanding of the sorts of refuges it utilizes. In the majority of instances, a reptile or amphibian will not be found simply lying out in the open. Normally, it will be partially or completely hidden from view by a structure or in a situation that is quite specific for that taxon. By analyzing habitat for the presence of key features, the likelihood of encountering reptiles and amphibians in pine habitats can be improved.

Forest Floor Litter

The surface of the ground in many tracts is quite barren and devoid of any cover. For some large and active species, this is an attractive environment and they can be seen out and about in direct view. However, the majority of pine wood herpetofauna are small and secretive, requiring microenvironments where they can keep out of sight of predators.

The least obvious hiding place is the most abundant. Except in the most xeric ridgetop sites, some sort of leaf litter usually blankets the forest floor. As naturalists move about the forest searching for features to look beneath to find hiding animals, they overlook the most important of these refuges as they trod across miles of it (plate 2.2).

Some layers of leaf litter are as dry as the sandy soil they cover and provide only visual protection. Oaks are the most prevalent deciduous trees in natural pine woods, and the leaf litter they produce is a favored microhabitat of small reptiles able to tolerate dry conditions. In tracts where populations of Mole Skinks, Crowned Snakes, Ringneck Snakes, and Short-tailed Snakes are present, individuals can be found hiding just below the surface of the dry, fallen oak leaves. Terrestrial amphibians require moist conditions while occupying areas away from breeding sites. In hilly areas, toward the bottom of gullies and ravines, fallen leaves accumulate and hold moisture, providing excellent conditions for Louisiana Slimy Salamanders, Mabee's Salamanders, Flatwoods Salamanders, and Striped Newts.

Fallen Logs

Fallen logs are a classic microhabitat for reptiles and amphibians. They retain moisture, both within their decaying, spongy wood, and in the soil over which they lie. In natural scrub areas they are rare because large trees are scarce. Abundant tree falls are most typical in untended forests or those managed exclusively for the betterment of wildlife habitat.

The only ecologically responsible way to search for reptiles and amphibians in and under fallen logs is by carefully rolling back the logs and then replacing them in their original position. Some species prefer to live beneath bark that has loosened on dead logs. These microhabitats are best left undisturbed, because there is no way to investigate the space underneath without destroying the environment that benefits herpetofauna (plate 2.3).

Animal Burrows

Although many endemic reptiles and amphibians spend much time underground, few dig permanent burrows themselves. Slimy Salamanders may on occasion create short tunnels where they stay for extended periods. Dusky Gopher Frogs construct permanent holes. However, the notion of a "snake hole" where some serpent has excavated a tunnel in which to live is a rare phenomenon in the real world of snakes, and no pine woods snake actually digs subterranean cavities in which to reside; a few species do use them for depositing eggs. Many species that utilize underground shelter do so by exploiting cavities and burrows dug by other creatures. Tunnels dug by crayfish in muddy areas bordering ponds provide a favored hiding place for Mabee's and Flatwoods salamanders. The importance of Gopher Tortoise burrows to Pine Snakes, Eastern Indigo Snakes, Eastern Diamondback Rattlesnakes, and Dusky Gopher Frogs is well known and described in the species account for the tortoise (plate 2.4).

One of the most unusual examples of a pine woods reptile or amphibian using the burrows of other animals is the proclivity of mole skinks and crowned snakes to hide within the pushed-up mounds made by tunneling pocket gophers. Crowned snakes prize pocket gopher mounds as basking sites. These snakes have a peculiar strategy for thermoregulating whereby they remain covered by sand with only their black-pigmented, efficiently heat-absorbing heads exposed. The gopher mounds provide optimal conditions for this behavior, as the gophers perform the activity in open, sunny sites, and the excavated dirt piles are loose and yielding to the efforts of tiny, burrowing snakes. Mole skinks find these mounds excellent sources of warmth during cool weather; they also use the piles of dirt excavated by certain burrowing scarab beetles as sheltering basking sites (plate 2.5).

Stump Holes

Hurricanes and tropical storms are an important shaper of habitat in coastal pine zones. While these storms often devastate human lives and property, they rarely cause lasting damage to wildlife habitat. In fact, some of the effects of fierce storm winds actually serve to improve habitats. One way storms do this is by downing trees. When a large pine tree is toppled, the roots are partially pulled from the ground, while some remain anchored. The cavities left by roots, and those that appear over time as still-buried dead roots decay, create deep tunnels used by many creatures to escape extreme temperatures or dry conditions. With most pine habitats devoid of rock outcroppings, decaying stump cavities are one of the few ways pineland herpetofauna can find subterranean refuge without having to excavate them themselves (plate 2.6).

Certain species are characteristically found around these structures. Black Pine Snakes frequent stump holes preferentially, and these holes have become even more important to the survival of these snakes as the Gopher Tortoise and its burrows have become increasingly scarce in the western portion of its range. Dusky Gopher Frogs, which like the Pine Snake are closely tied to tortoise burrows, are relying more heavily on stump-hole shelters in De Soto National Forest. Slowinski's Corn Snakes often hibernate in stump holes and are easiest to find in the early spring as they emerge from winter dormancy and lie near them.

Artificial Microhabitats

Manmade structures and modifications of the land sometimes create favorable conditions for herpetofauna. One thing most pleasing to a field herpetologist is a pile of old boards, discarded appliances, and, most cherished of all, sheets of corrugated roofing tin scattered about in partial sun along some rural roadway. Except during very cold or hot conditions, such trash piles are usually gold mines for a biologist. The moist, protected space beneath these objects is extremely attractive to reptiles and some amphibians (plate 2.7).

Corrugated metal used for barn and shed roofs often ends up on the ground at the site of dilapidating structures. It absorbs and retains heat, and in the Deep South, where winters are comparatively mild, the warmth under a sheet of tin is sufficient to keep many snakes up on the surface in December and January. Later in the spring, trash piles are also very productive sites, and checking them early in the morning before some of the residents warm up and disperse is the best time to make exciting discoveries (plate 2.8).

Most amphibians require transient bodies of standing, fish-free water in which to complete their life cycles. Such environments are rare in savannas and, especially, ridge associations. It is ironic that one of the degrading activities that people inflict on these lands—the construction of homes and commercial buildings—can create useful features for native amphibians. When construction sites require large amounts of earth to elevate a site or level a grade, it will often be obtained by digging it up from a nearby location, leaving a depression. Some of these "borrow pits" hold accumulated rainfall long enough to serve as breeding and larval development sites for a variety of species, including Mabee's Salamander and the Pine Woods Treefrog (plate 2.9).

Ecotones are narrow spaces where two distinct habitats abut one another. One of the most starkly contrasting ecotonal junctions occurs where clearcuts border stands of mature timber. This situation is a natural occurrence on a small scale when tree falls from storm damage or disease are present, creating a forest gap. Gaps often display their own unique ecology and faunal and floral communities. In commercial tree farms, ecotones bordering clearcuts are ubiquitous and extensive. They are favored locations for some species of snakes and lizards if the general area is suitable. Assuming the taxa are present in the vicinity, Brownchin and Tan Racers, Eastern Coachwhips, and Mimic Glass Lizards will utilize these harsh transitions between sunny and shaded environments (plate 2.10).

3

Pine Woods Generalists

Louisiana Slimy Salamander
Plethodon kisatchie

Of all the endemic herpetofaunal elements of the pine woods, a lungless salamander is the most surprising. In several important ways, the biology and physiology of the Louisiana Slimy Salamander strays from the norm in taxa adapted to this peculiar environment. In fact, the Slimy Salamander gives the impression of being a species constrained, even trapped, by the conditions characterizing pine woods habitats, and contrasts with the other herpetofaunal elements that in observable ways have adapted to and integrated themselves into these parameters.

Viewing a topographic map of the parishes of north central Louisiana reveals a hilly land, dissected by creeks and drainages. These features define an environment noticeably different from the surrounding country, for along these low-lying areas are shady, mesic microenvironments. Soils here are more organic than in the sandhills that encompass them. Any herpetologist would recognize these sites as salamander habitat.

Perhaps it is inaccurate to consider the Louisiana Slimy Salamander a true pine habitat resident, because it shuns the xeric ridge and savanna habitats that occur throughout north central Louisiana, inhabiting instead the creekside and ravine bottomlands that make feeble incursions into these dry territories. In essence, *Plethodon kisatchie* does not live in the southern pine woods, it is held captive by them. It survives in scattered parcels of typical lungless salamander microhabitats that exist like small islands within a broad sea of dry pineland.

Description

Louisiana Slimy Salamanders are elongated creatures with long tails and reduced limbs. They are black or deep bluish black over the dorsal surface. Their venters are barely lighter than the dorsums, being a dark smoky gray, and this color is often seen on the lower sides of the tail. The entire dorsum is suffused with silvery white flecks as if sprinkled with tiny flakes of dried white paint (plate 3.1).

The most notable physical feature of the Louisiana Slimy Salamander, as with others so named, is the amazingly sticky secretion it produces when handled. Within moments after picking up this creature, the hands become smeared with a gluelike substance that adheres to skin like virtually no other material in nature. No amount of soap and water seems to be able to remove it. Vigorous rubbing with a coarse, dry cloth is the most effective way of lessening the sensation of sticky gum on one's hands. Time and wear alone are the elements to which the salamander's slime eventually yields.

No morphological character distinguishes the Louisiana Slimy Salamander from other species within the complex. The phenomenon of cryptic species and the underestimation of biodiversity they create is a prominent issue in the consideration of pine woods herpetofauna. Including the present species, biologists have overlooked four taxa endemic to pineland ecosystems, at least in terms of their taxonomic significance if not of their existence. Mimic Glass Lizards, *Ophisaurus mimicus*, were handled by herpetologists in the South for years but passed off as aberrant Slender Glass Lizards, *O. attenuatus*. The unusual coloration of the Corn Snakes in the pine hills of central Louisiana was common knowledge among herpetologists for decades, but only in the last few years have they been appreciated as a definable taxon. Since its description in 1929, the Louisiana Pine Snake has been considered a rare subspecies of a generally common species of North American snake until very recently. The Tan Racer is another cryptic taxon, for it was only a thorough 1970 study of racers across a large geographic area that put the uniquely colored snakes of southwest Louisiana and east Texas into proper perspective. The Louisiana Slimy Salamander is the most cryptic of the pine woods herpetofauna, being completely indistinguishable from related species, even to specialists, without the technical means of visualizing genotype. The fact that this species was recognized only after many decades of being shrouded from view by its superficial resemblance to other salamanders serves to caution biologists that we may still be underestimating the extent of remarkable animal forms that have evolved exclusively in particular habitats. Perhaps the full list of pine woods herpetofauna has yet to be compiled, and organisms that are already within our view embody these future taxonomic designations.

Distribution Notes

Because of its requirement for moist habitat within an environment not noted for this quality, the Louisiana Slimy Salamander has a patchy occurrence within its overall distribution. The notable distinction of the landscape where it occurs is the hilly relief of the land. In Louisiana, records for *P. kisatchie* do not stray outside the north central region sometimes referred to as the hill country. This area has abundant creeks and small streams that run along ravines and create favorable conditions for woodland salamanders. Its distribution in Louisiana encompasses 10 parishes in the north central portion of the state. Like the Louisiana Pine Snake, its disjunct distribution suggests it is a relict from a time when its ancestors ranged continuously across the west gulf coastal plain. Isolation was caused by the development of the Mississippi River Delta region to the east and, to the west, a receding pine forest biozone in east Texas.

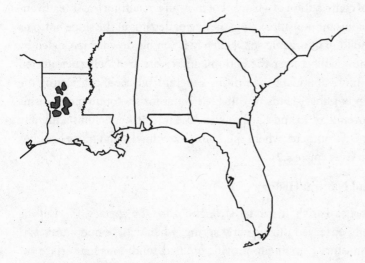

Habitat

Plethodontid salamanders are aptly known as lungless salamanders. Respiration takes place directly with the outside environment via gas exchange across the cutaneous membrane. This mechanism is dependent on the exchange membrane remaining perpetually moist. Therefore, it is no surprise that plethodontids exhibit greatest species variety in the mist-laden Appalachian Mountains, with their abundance of wet and shaded ravine and slope microhabitats. What a stark contrast to this optimal habitat are the seemingly inhospitable trans-Mississippi sandhills! Yet *P. kisatchie* strays but little from its familial habitat preference, and thus is relegated to small and scattered sites embedded within the matrix of drier pineland.

Because of their need to maintain a moist skin, lungless salamanders are typically found in mesic microenvironments, directly in contact with a wet substrate. Specimens are often discovered when a fallen log, rotting and mushy with absorbed water, or a rock resting atop soggy ground, is lifted to reveal the hidden occupant. The same can be said for the Louisiana Slimy Salamander, the difference being that finding such objects to look under can be quite difficult within this species' distribution. The ridges and bottoms that alternate in sequence across broad swaths of the region are predominantly sandy, and they drain so quickly that lifting most objects will uncover ground too dry for any lungless salamander to occupy. Only in low ravines that function as intermittent creeks does enough moisture occur with any consistency, leading to the development of shrubs and deciduous trees and their annual contribution of a blanket of moisture-retaining leaf litter. Eventually, this constant moisture and gathering of detritus at the slope bottoms leads to a buildup of organic soil. Fallen logs tumble down slope, aided by rain and snow, collect near the bottom, and create areas where protected refugia are more abundant than in the general landscape, completing the process of providing islands of suitable habitat for mesophilic organisms within an overall xeric and harsh environment. It is here, amid a literally suffocating environment, where the Louisiana Slimy Salamander breathes easy and survives (plate 3.2).

Ecology and Natural History

In Natchitoches Parish, a habitat stronghold for the species, *P. kisatchie* is most easily observed during early spring, when cool temperatures and abundant moisture combine to provide optimal conditions for surface activity. Otherwise, the habitat being rather harsh for salamanders in general, *P. kistachie* is elusive. Specimens are usually found by moving fallen pine logs lying along ravines and especially at the bottom of these slopes, where ground debris accumulates and drainage is poor. During the summer, Louisiana Slimy Salamanders occasionally turn up during periods of prolonged and heavy rainfall. During most of the summer, the salamanders move deeper underground and become more sedentary. At these times, they are very difficult to find.

Slimy salamanders feed on invertebrates and are not selective as to type, consuming a variety of invertebrates small enough to be overcome and swallowed (Pope 1950). In forests where management includes the application of herbicides and pesticides, Louisiana Slimy Salamander populations may be negatively impacted by a reduction in available prey.

A multifaceted courtship precedes mating. During the breeding season, males respond to the close proximity of another Slimy Salamander, regardless of sex, with a series of stylized leg movements. If the individual of interest is a female, the response will identify her as such to the male, and courtship proceeds. However, some males take advantage of this fact, and mimic the behavior of females (Arnold 1977). If the ruse is well executed, the courting male is fooled throughout the entire courtship process and may deposit his spermatophore for the charading male, thus wasting a limited ability to fertilize any females in the vicinity, to the potential fitness advantage of the male masquerading as a female.

A courting male moves into close proximity to a female and trails his rostrum along her body, focusing his contact point on his mental gland under his chin. Quickly thereafter begins a phase involving stiffly raising and lowering the legs, first the hind legs, then the front legs, in a sort of step-dance that continues until the next stage of courtship begins (Petranka 1998). Next, the male positions himself so that his tail is presented to the female. He accomplishes this by slowly moving up the female toward her head and then deftly slipping beneath her while continuing to advance forward until she is situated directly above his tail, either straddling it or lying alongside it. The male then begins to wave his tail from side to side, which attracts her attention to it. The goal is to guide the female to straddle his tail and follow him on a lengthy march across the ground. If the female does so, he will lead her until he finds a suitable location for spermatophore deposition—a dark, protected, and damp situation. Upon deposition of the spermatophore, the male, still leading the female by his tail, brings her to stand directly over the structure whereupon she grasps it with the marginal lips of her cloaca.

Plethodon kisatchie females oviposit during the summer. In typical fashion for the species complex, females secrete themselves in cavities under or within rotting logs where they deposit their clutch. The egg mass is an adherent gelatinous mass hung from the ceiling of the nesting cavity. Female salamanders remain with their eggs for the duration of the 2–3-month larval period, and are probably effective in defending the clutch from invertebrate predators. Maternal care may extend beyond the larval period in some cases, but observations are spotty in general and nonexistent for *P. kisatchie*.

Conservation

Since this salamander is not specifically adapted to the xeric conditions of pine woods but, rather, trapped within them in embedded areas of suitably moist habitat, its actual area of occurrence is far more spotty and disjunct than can be represented on the range map. Even at a relatively local scale, the salamander will be alternately absent and abundant as one surveys the dry ridges spaced between more mesic ravines and creekside zones. Adding to its heterogeneous population distribution is its apparent avoidance of cultivated tree farms. Perhaps because of the paucity of fallen limbs and vegetation cover other than pine straw, the Louisiana Slimy Salamander is not to be expected in commercial stands of pine where fire is suppressed. Extensive search for the species across its distribution has borne out the hypothesis that the Louisiana Slimy Salamander thrives only in mesic microhabitat within relatively undisturbed pine woods.

Summary

The other two species of salamanders peculiar to the southern pine woods—the Flatwoods and Mabee's Salamander—are better able to function in a relatively dry environment by being independent from exclusively cutaneous respiration. Furthermore, they are more adept at escaping truly xeric conditions that exceed even their tolerance limits by their fossorial tendency, exploiting subterranean microhabitats where some moisture is retained. Although the Louisiana Slimy Salamander is equally secretive, it is more typical to be uncovered within the recesses of a rotten log than below the substrate surface. Given that such refuges are uncommon in the southern pine woods, and most fallen limbs and logs are dry to their pith for most of the year, the Slimy Salamander is heterogeneously distributed. *Plethodon kisatchie* is less likely to occupy such situations than the two species of *Ambystoma*, and on balance seems less a product of adaptation to pine woods than a survivor in an unfavorable environment. Rather than adapt to an environment not typically occupied by members of its family, this species survives in its broader habitat while maintaining a dependence on the limited areas where benign conditions manage to persist.

Pine Woods Treefrog
Hyla femoralis

Many pine woods species require special effort to see because they rarely stray from the optimal natural conditions that compose their microhabitat. A Mabee's Salamander may best be sought by driving to a rural site and traversing into piney wetlands, and the hunt usually succeeds only after a lengthy session of logrolling. To find a Louisiana Pine Snake, even a resident of one of the small towns within the general range might need to hop into a sturdy four-wheel-drive vehicle and follow sand tracks deep into ridgeland scrub for even a slim prospect of success. These days, the few people who ever knowingly encounter a Mimic Glass Lizard are biologists who go to considerable trouble to be situated in the right location, or people whose livelihood compels them to leave the comfort of town, home, vehicle, and road to spend hours at a time in areas where no indication of the hand of man is apparent. Yet, among the herpetofauna of the southern pine woods, there are a few opportunists. This small cadre of species—Slowinski's Corn Snake and Key Ringneck Snake for example—has been able to utilize the alterations that humans have made on their habitat and are as likely to be found in a backyard as on a wilderness tract.

The capacity of some reptile and amphibians to survive the push of human development is attributable to a variety of reasons. Slowinski's Corn Snake is a true ecological generalist with a broad palette of attributes that enable it to live equally well under a range of conditions. The Key Ringneck Snake owes its survival in residential areas to the extremely narrow scope of its highly specialized needs. The ringneck's reliance on only a few square meters of mesic, shaded soil is admirably provided by a fragment of plywood lying in a carport, even though the snake would naturally be living in a secluded corner of a pineland wilderness.

The most adaptable and conspicuous of pine woods herpetofauna is the Pine Woods Treefrog. Within the southern pine belt, provided that a semblance of pine woods lies within reasonable proximity, this frog is as productively sought in cities and towns as in pristine habitat. It is a species that has been shaped and defined by the pine woods, but seems less firmly tied to them than most other herpetofaunal components. It is a common sight on humid nights as it clings to windows and walls of shops in the downtown districts of towns along the Gulf Coast. After dark, around street lights and backyard porch lights where insects congregate, one to several Pine Woods Treefrogs will often be found exploiting the serendipitous buffet.

I noticed the largest aggregation of Pine Woods Treefrogs I'd ever seen while waiting out a heavy summer thunderstorm under the awning of a convenience store on the outskirts of Pensacola, Florida. I happened to glance behind the ice machine that sat just outside the store entrance. There, between the building and the machine, was a congregation of about a dozen *Hyla femoralis* along with several Green Treefrogs, *H. cinerea*. The passing storm had caused the frogs to rouse and shuffle their positions, attracting my attention. A swampy tract running behind the stores had undoubtedly been pine woods before newly built suburbs created economic support for the array of new businesses that had sprung up along the highway. The flooded ditches and ratty swaths of pine trees were enough to maintain a population of Pine Woods Treefrogs, the most resilient and persistent of the pineland amphibians.

Description

Hyla femoralis bears a pattern and coloration that impart excellent camouflage when the frog is perched on a pine tree. The dorsum is marked with a dark blotch that varies in shape among individuals. Most often the marking has irregular extensions extending outward as "arms" making it resemble a misshapen X or a splat of dark paint. On some specimens the marking is broken into two or more separate blotches. The dorsal pattern is usually distinct, but as with most *Hyla* (but excluding *H. andersonii*), *H. femoralis* can change its hue significantly, depending on temperature, light level, or activity of the frog. When a Pine Woods Treefrog is at the palest or darkest extremes of its coloration range, the characteristic dorsal marking can be difficult to see (Plate 3.3).

Another identifier is the color of the inner thigh. Several *Hyla* of the eastern United States have bright colors on the hidden portions of their bodies, and these aspects are not prone to vary in color or intensity when the dorsal coloration changes. In *H. femoralis*, the portions of the thighs not seen when the frog is in repose are marked with bright yellow or orange spots (Mount 1975).

Hyla femoralis is a medium-sized treefrog, with an average adult SVL (snout to vent length) of 40 mm (Mount 1975).

Hyla femoralis is karotypically unique among North American hylid frogs. It is the only species known to possess heteromorphic sex chromosomes, of the XX/XY type (Wiley 2003). Furthermore, it exhibits a metacentric chromosome 6, also apparently unique among regional hylids (Wiley 2003). Through comparison of karotypes, Wiley further proposed that *H. femoralis* was derived from the Gray Treefrog, *H. chrysoscelis*.

Larval *H. femoralis* are quite spectacular. Full grown, the tadpoles measure 20–40 mm. The most prominent features are the high dorsal and ventral caudal fins, boldly marked with red and black. The dominant color of these broad tail fins is deep blood red to bright orange, with broken black borders at the edge of the fins and fainter black marks within the red areas. The tail musculature is marked with a clean white stripe running longitudinally, bordered with black (Gregoire 2005).

Distribution Notes

This frog is the widest ranging of any of the pine woods endemic herpetofauna. In terms of extensiveness of distribution, it is closely approached only by the Oak Toad, but that species is spottier in occurrence, being more habitat specific. *Hyla femoralis* occurs throughout the entirety of the historical distribution of longleaf pine with one caveat. The species has not been recorded in western or central Louisiana, nor in extreme eastern Texas. Thus, even this far-ranging amphibian does not violate the biogeographical

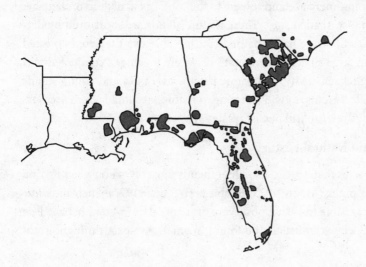

boundary of the trans-Mississippi pine woods. However, in its near pande-mism across the entirety of the southern pine woods, *H. femoralis* is unique among the taxa covered by this book. It inhabits many offshore islets and sandbars so long as intermittent freshwater pools are present (Irwin and Irwin 2002). As a generalist species, it is likely to be harbored in any location grown to pine and offering ephemeral pools during the breeding season, from southeastern North Carolina to the Louisiana Florida Parishes.

Habitat

Hyla femoralis might well inhabit all southern pine communities if not for its need for arboreal shelter. Some natural scrub has canopy coverage barely exceeding 10 percent. In such exposed terrain, the few trees that are present are stunted, resulting in arboreal refuges too hot and dry to be utilized by the Pine Woods Treefrog. In plantations, canopy coverage can range from dense to absent, and a shaded mesic tract can be transformed in a day to a barren, hot landscape. Thus, *H. femoralis* may be abundant, scarce, or absent in plantation tracts within the same general area. In other situations, from wet flatwoods to high, dry ridges, *H. femoralis* is a significant component of the herpetofauna. *Hyla femoralis* requires secure and mesic arboreal mi-crohabitats such as loose bark clinging on trunks, tree forks with accumula-tions of pine straw, and fissures in dead snags.

Young, even-aged stands of pine, such as seen in successional patches after storm or severe fire damage, or in areas replanted after logging, are un-suitable for this amphibian. Stands of mature saw timber offer the best habi-tat in many commercial stands owing to the component of dead or damaged trees and moist, shaded sites. These mature stands are short lived by defi-nition—once the trees have attained considerable size they are harvested. However, in large silviculture operations where logging is conducted with environmental sensitivity, a dynamic patchwork of old and young stands, interspersed clearcuts notwithstanding, is maintained and in such areas the Pine Woods Treefrog will persist (plate 3.4).

Ecology and Natural History

This is a species that, true to its name, spends most of its nonbreeding time high in the pines (Mount 1975; Babbit and Tanner 1997). Such individu-als are inaccessible to casual observers of nature, but in good habitat Pine Woods Treefrogs can usually be found in small numbers throughout the year under fallen logs or in damp places under loose pine bark (Babbit and Tanner 1997).

The reproductive season is lengthy, with choruses sounding at the onset of hot, rainy weather anytime during spring or summer. The call is a rapid "kek-kek-kek." Clutches are usually 100 or so eggs, laid in clumps and adhering to emergent vegetation, although some eggs may also be left to float at the surface of the pond. Larvae metamorphose in 6–10 weeks.

Conservation

Hyla femoralis is not a species in jeopardy. It is adaptable to a variety of situations common throughout the southeastern pine belt, and is abundant near homes and within towns and small cities. Vulnerabilities are its apparent need for some degree of cover in trees of moderate size and, like so many amphibians, its requirement for transient rain pools unpolluted by toxic runoff. Babbit and Tanner (1997) commented that landowners taking responsibility for maintaining suitable habitat for this species should preserve timbered tracts. So long as extensive clearcutting is avoided and herbicides and pesticides are not broadcast or used carelessly near pond basins, even commercial tree farms can provide refuge for *H. femoralis*.

Summary

The Pine Woods Treefrog is a species that is widespread and abundant. The frog is familiar to herpetologists, but fairly general statements make up the bulk of literature on the species, with many gaps in our knowledge of its habits. With time, broad general statements attain weight, and the impression one can get is that there is very little left to discover about the species. Thus, we know it to be a "summer breeder" stimulated by "heavy, warm rains" and, at other times, a generalist invertivore residing high in the pine canopy. It remains a paradox that one of the most common members of the pine woods herpetofauna is one of the least studied.

Oak Toad
Anaxyrus quercicus

When Hurricane Ivan roared onto the Alabama coast in 2004, it caused a disaster that destroyed human lives and livelihoods and laid waste to the businesses and summer homes in and around Gulf Shores and Orange Beach. Ivan forever altered many people's lives, and from the human perspective, the best thing that could be said about the damage inflicted was that some features of prestorm life would eventually return, in spite of, rather than because of, Ivan. What a contrast then, was the impact of the hurricane on the natural ecosystems and its consequences for many fauna of the coastal pine woods. Among the leaf-strewn sandy flats preserved within the Bon Secour National Wildlife Refuge live thriving populations of Oak Toads. On that September day, the awesome winds and catastrophic rainfall inflicting terror on the human residents of Gulf Shores were simultaneously pummeling the home of the Oak Toad. Compared with the devastation inflicted upon the citizens of Gulf Shores, the effect of Hurricane Ivan on such a delicate-looking creature was amazingly benign, even beneficial.

With much of the historical stands of southern pine woods situated near the coastline, tropical storms have been sporadic but consistent forces in shaping these woodlands. Like fire, storm-wrought wind and water disturbance has impacted the wildlife residents for centuries. Yet the presence of a complete herpetofauna in even the most vulnerable coastal and insular sites attests to the remarkable resilience of these animals to the effects of hurricane strikes. Following significant restructuring of their habitat by events such as storm surge overwash, vegetation dieback due to sea spray, and extensive tree falls, reptiles and amphibians can still be found, mere weeks later, leading apparently undisrupted lives.

Species that are normally very conspicuous illustrate this phenomenon more clearly than species that are difficult to observe under any circumstance. Oak Toads are among the most obvious elements of the southern pine woods herpetofauna. Little is known of any territorial tendency in this species, and they certainly appear to accept living in close proximity to each other and are often found in high densities. They are decidedly diurnal, rather than nocturnal, unlike almost all of their native brethren, so if a habitat supports them it is usually readily apparent to even a casual human observer. Following the devastation left by a hurricane such as Ivan, one might expect that a stretch of coastal pineland that once harbored a population of tiny toads would be devoid of them for many years to come.

On my first visit to the Bon Secour National Wildlife Refuge in the summer of 2004, I spent a few days devoted to observing reptiles and amphibians in the Little Point Clear Unit, which encompasses the most xeric habitats of the Refuge. With no place in this tract more than 4 km from the sea, and its low relief, this is an area that would receive the full brunt of any passing tropical storm. Little did I know that in just two months, the eye of a Category 3 hurricane would pass directly over the flatwoods I was exploring. During my quick survey I found large expanses of palmetto flatwoods intermingled with savannas atop slightly raised platforms of land. The trees were predominantly a mix of mature slash pine and oaks, with occasional ancient-looking magnolias here and there, and the general character of these woods was of rather dense flatwoods forest. The site presented itself as ideal habitat for Oak Toads, with a heterogeneous mix of wet and dry pine woods and abundant oak leaf litter and fallen limbs. As I anticipated, toads were everywhere abundant, and I encountered one every minute or so while walking through the forest. Most specimens I saw were subadults and appeared to be of roughly the same cohort, but a sizable number of full-grown adults were also out and about.

Following Ivan's passage, I made a return trip to Bon Secour to assess the damage to the herpetofauna and their habitat. The first thing that struck me when I entered the refuge was the relatively slight impact of any type that could be discerned. I had been preparing myself for a scene of massive destruction, half expecting to see the forest completely flattened and the terrain scoured of most of its vegetation. Certainly the drive through the City of Gulf Shores, on the way to the refuge, bolstered this fear. Every structure on or near the beach had taken some form of damage, and many were completely destroyed. Some large and sturdy looking homes I remembered from my previous trip were simply gone altogether, unmoored from their foundations, swept across the barrier island, and dumped into the backwater bay on the other side. In stark contrast, walking through the forest gave no such sense of the recent landfall of an awesome storm. The basic landscape stood unscathed, and the broad vistas before me appeared but slightly altered and much as I remembered them from earlier in the year.

Closer consideration revealed a few changes. Stands of pine immediately adjacent to the coastal dunes had been afflicted by the continuous salt spray and ocean overwash past the beach zone, and most had dead brown crowns. Most of the larger pines nonetheless indicated their tenacity by sporting tufts of new-growth green needles at the ends of some of their branches, previewing a rapid recovery of these stands before the summer had passed. The young sapling pines were less able to survive the invasion of salt water and most had died, but this had simply opened up the midstory layer and new seedlings would soon replace them.

Although most of the trees remained standing after the storm, some had been destroyed. Two basic types of damage occurred. Some trees, particularly the slash pines, were twisted and torn by the high winds and snapped off above the ground, leaving behind a standing trunk base with its broken end frayed like heavy twine. Many of the largest oaks were pushed over and lay prone on the ground. Where such trees had once been planted, there was now a shallow depression left by the roots as they lifted large amounts of earth with them.

These effects of Hurricane Ivan on the Bon Secour flatwoods might easily have been considered damage by a human assessor. From the perspective of the Oak Toad, these were merely modifications either of little consequence or with beneficial effect. The salt burns that had killed the smaller slash pines would impart a beneficial long-term effect on these woods. Already I could see indications that the grasses and herbaceous ground cover had not been extirpated and were beginning to sprout in patches. The loss of small pines had opened up the habitat and the increased sun exposure would accelerate the recovery of the understory when the warm spring rains returned. These rains would wash any remaining salt out of the surface microenvironments in short order. The net effect of the salt water intrusion would be similar to that of a fire, cleaning out excess vegetation and maintaining the character of a scrubby flatwoods environment to which the local herpetofauna was well adapted.

The small basins created by the toppling of trees would function as catchments for rainfall, and, indeed, some of these basins were already functioning as small ephemeral pools. The deepest of these would hold water long enough to serve as additional sites for breeding and metamorphosis and were a harbinger of a bumper crop of Oak Toads on the near horizon. Even the more superficial depressions would provide convenient refuge for toads because they would fill briefly during rains and remain damp longer than the surrounding earth. The fallen trees would likewise increase the availability of normally scarce refuges; decaying logs hold moisture within and beneath them and are a favorite hiding place of amphibians, but are usually less abundant in sparsely treed pine woods than in the more densely canopied deciduous woodland.

It was under one such recently felled slash pine that I discovered a survivor of the storm. A large female toad was nestled in a form-fitting depression she had fashioned beneath the tree trunk. In a few months she would be lured by the calling males and would soon recharge this corner of the

Bon Secour refuge with 100 or more little toadlets. The toads had survived Hurricane Ivan, just as they had many unnamed and unrecorded hurricanes over the centuries.

Description

This is an elfin toad bearing a distinctive pattern. Size alone is diagnostic, as the Oak Toad is the smallest member of the genus in the United States. Adult SVL is 19–32 mm (Conant and Collins 1991), with females larger on average than males.

The pattern consists of 4–5 pairs of dark black spots on either side of the midline, divided by a thin, bright stripe that is white, yellow, or orange. Dispersed abundantly over the dorsal surface are dark red or orange tubercles. Occasionally one will find melanistic specimens with no discernible pattern. The venter is grayish white. The limbs bear black bars on a gray base color (plate 3.5).

Like the adults, larvae are marked with a series of 6 or 7 dark dorsal markings. The dorsal half of the tail is heavily marked.

Distribution Notes

A wide-ranging species with essentially the same distribution as *Hyla femoralis*. Both species are the only habitat endemics that occur virtually throughout the southern pine belt. The Oak Toad ranges across the entirety of the Atlantic coastal plain south of central Virginia, thus including the small portion of extralimital pine habitat in southeastern Virginia, as seen

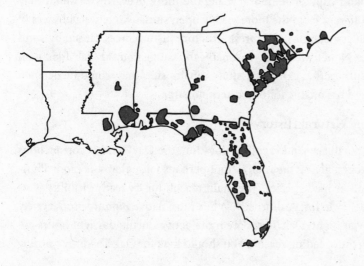

with the Pine Woods Treefrog and Mabee's Salamander. Historically, the species occurred everywhere in Florida, including the Keys, although large gaps in its distribution created by development are now present. The range continues along the gulf coastal plain and terminates in the Florida Parishes of southeastern Louisiana.

Habitat

Anaxyrus quercicus occupies a range of habitats, from flatwoods to xeric ridge associations. Their abundance seems incongruous in high pine, where midstory vegetation is often dominated by stunted oaks; hence their common name is a reminder of just how remarkable and distinctive these little toads really are. Almost all pine woods amphibians exhibit bimodal habitat occupancy; adults return to relatively dry landscapes following breeding events in wetlands and typically become inactive while away from surface water. The Oak Toad is a species that not only occupies dry ridges, it continues to be a fully active, diurnal forager in this environment, and in this respect it is unique among its amphibian sympatrics.

Anaxyrus quercicus is also tolerant of variations in fire regime. Greenberg and Tanner (2005) found that among their study populations, those with lengthy fire suppression histories that allowed hardwood midstory development fared better than nearby populations living on fire-maintained savanna. This suggests an overall larger population of toads in overgrown pine woods and differs from conventional wisdom for the conservation of most inhabitants of pine habitats. Greenberg and Tanner cautioned that in a broader landscape or temporal scale, the more overgrown habitat may not be so different from the more visibly open sites, owing to patchiness in vegetational associations and forest structure disturbances caused by wind and wildfire. Nonetheless, it is reasonable to assume that the Oak Toad is an adaptable species that can successfully inhabit sites that would be deemed degraded and unsuitable for other herpetofauna.

Ecology and Natural History

Despite its small size and cryptic coloration, the Oak Toad is conspicuous where it occurs, giving the impression, perhaps false, that it is more abundant than other toad species. The main reason for its ready visibility is its strong tendency to forage diurnally. While most native *Bufo* are productively sought only at night, Oak Toads are more active during daylight hours except during the breeding season. You should look for Oak Toads by walking slowly through appropriate habitat on bright sunny mornings, particularly after a previous night's rainfall, pausing every few steps and scanning the ground for movement and listening for the faint sound of the little amphibians as they hop away across fallen pine needles and oak leaves.

Oak Toads breed throughout the spring, summer, and early fall. They are stimulated to call by warm rains and will begin moving toward spawning sites after several successive days of such conditions, or following single weather events that bring very heavy precipitation. Any transient body of water may be utilized, ranging from natural depression marshes to roadside ditches. Even rain filled ruts made by vehicles on primitive trails will serve as a breeding site for *A. quercicus*.

Oak Toads have sun compass orientation, that is, they use the position of the sun to orient themselves in the landscape and to find their way into breeding ponds and back to their home territories after breeding (Goodyear 1971). In such a harsh environment as typical pineland situations, it is beneficial to be able to return to locations that have previously represented adequate hydroperiods. Amphibians often return to sites with proven prey base and good microhabitats rather than wander into unknown territory in a randomly chosen direction.

The call of *A. quercicus* is quite different from that of other U.S. toads. The sound is best described as a "peep-peep-peep," very sweetly toned and high pitched. It has been compared to the sound of the furtive cheeping of a baby chicken. Wright and Wright (1949: 198) commented that the sound of a calling Oak Toad was "the most unfroglike note I ever heard."

The eggs are sometimes laid singly or, more often, in short strings that contain 2–8 eggs. They are usually attached to vegetation that lies closer to the bottom of the pond than the surface. Fecundity is quite low compared with that of other southern bufonids, from 100 to 250 eggs per female being typical.

Larval development is rapid, and newly metamorphosed toads are 7–8 mm in SVL (Hamilton 1955). Subsequent growth and maturation is also rapid in this species. Hamilton (1955) measured individuals recaptured during one spring to fall period on a site near Ft. Myers, Florida and reported an average growth rate in that year's recruits of 2–4 mm SVL per month. At that rate, and given the nearly 10-month activity period these toads experience in south Florida, Hamilton estimated that males mature in time to breed in the first spring following their larval period, and that the larger females probably require two years to attain sexual maturity. However,

Greenberg and Tanner (2005) conducted a long-term study of a metapopulation in central Florida and determined that most individuals of both sexes return to breed the year following metamorphosis for their only reproduction. They were presumed dead two years after metamorphosis because they were never trapped after the first breeding event. These researchers detected a small cohort of toads that survived for two to four years after metamorphosis, but these were individuals not known to have migrated to breeding sites during any previous season, suggesting that breeding immigrations and emigrations and time spent in the ponds is a significant contributor to the apparently high annual rate of mortality in some *A. quercicus* populations.

Dry, porous soils in the pine woods contribute to erratic availability of suitable spawning sites, and Oak Toad populations may experience several years with no reproductive recruitment. A body of theory holds that many amphibian taxa persist through time through metapopulation dynamics (Semlitsch et al. 1996). Under this model, local population extinctions caused by a variety of factors are quite the norm. The loss of local groupings is offset by the dynamics of immigration from neighboring populations into decimated sites and emigration of some individuals in dying populations out to new locations of suitable habitat. In this way, the species maintains its overall distribution and abundance. Greenberg and Tanner (2005) monitored eight populations of *A. quercicus* through ten consecutive years and observed significant juvenile recruitment during only two seasons, involving just three of the populations. However, these toad populations did not readily fulfill predictions of the metapopulation dynamics concept, with few instances of movement between neighboring ponds having been observed and no population extinctions despite very long periods with no recruitment. Viewed from the perspective of one concerned with the conservation and preservation of amphibians, Greenberg and Tanner's study offers hope that at least some taxa are more resilient and steadfast than often assumed.

Oak Toad population density can be quite high, but localized. Hamilton (1955) recorded collecting 49 *A. quercicus* in an hour while walking along a roadside ditch that held no water. On another site he estimated a density of 173 per ha, but noted that a similar site only 6.4 km distant was without any visible toads. Calling the abundance of Oak Toads "incredible," this aging study describes a historical environment that contrasts with most modern-day experiences in Florida pine woods.

The home range of *A. quercicus* is quite small. Hamilton (1955) recaptured marked individuals over the course of a month. His data revealed that the toads generally remained within 30 meters of their initial point of capture. The home range of these individuals averaged roughly 15–20 square meters.

Hamilton also presented data regarding the dietary preferences of this species. During the brief sampling period, the most abundant prey item identified was ants. While it is possible that the components of various prey in the diet of Oak Toads changes as invertebrate abundances shift through the seasons, ants would be readily available throughout the year. Thus the Oak Toad may play a specialized role as an ant predator compared with other sympatric anurans.

Conservation

No evidence exists to indicate the Oak Toad has lost significant areas of habitation or major populations. Carr (1940, quoted in Hamilton 1955) declared that "Today I believe it would be possible to cross central Florida by car and at no time be in a situation where the calls of *quercicus* were not audible." Hamilton (1955: 205) described the species as "extraordinarily abundant in much of its range." While continuous abundance across large swaths of territory is not the case today, certainly for such a wide-ranging amphibian ample strongholds can be found where the toad is as abundant as it ever was. With the present fragmented state of undeveloped areas, however, there are locations where the habitat has been altered to the point of reducing or eliminating local populations. With less than 3 percent of the historic longleaf pine habitats still extant in the southeastern United States, and other types of pine woods under assault as well, no southern pine woods endemic can be considered secure.

Anaxyrus quercicus, being an adaptable habitat generalist, is currently faring well within its changing environment. Problems may be on the horizon, however. Limited data suggest that *A. quercicus* is a specialist feeder on ants. The notion of an ant predator living in the southeastern United States immediately causes concern, because such a species would be especially vulnerable to mortality from attacks by Red Imported Fire Ants, *Solenopsis invicta*, which are on an out-of-control ecological rampage across the region.

Summary

Within the context of the southern pine woods ecosystems, the Oak Toad is a habitat generalist. It is interesting to compare the Oak Toad with a habitat-specific endemic like the Dusky Gopher Frog, and the differences in their wild status. The toad, although less common in some locations than in times past, is still found in all states and general areas where it has been historically documented. The frog, once known from southeastern Louisiana, southern Mississippi, and southwestern Alabama, has disappeared from Louisiana and Alabama and is now reduced to a few tiny patches of forest in southeastern Mississippi. While the reasons causing the decline of the frog are complex and not completely understood, one of its vulnerabilities is its dependence on narrow habitat parameters. The Oak Toad, in sharp contrast, is just as noteworthy for the wide variety of situations in which it is found.

Slowinski's Corn Snake
Pantherophis slowinskii

As the southern pine woods have undergone a dramatic transition toward fragmentation and domestication, most endemic reptiles and amphibians have, at best, managed to endure. Perhaps only one of them, Slowinski's Corn Snake, has been able to exploit the changes with such efficacy that its abundance and distribution, by any available measure, has remained unchanged. Since this snake has been recognized as a distinct taxon only since 2002, very little attention has been directed its way and descriptions of its historical abundance do not exist. Nonetheless, in every way, Slowinski's Corn Snake resembles its more familiar sister taxa (Smith et al. 1994), the Eastern Corn Snake, *Pantherophis guttatus*, and the Great Plains Rat Snake, *P. emoryi*, both well-studied serpents known to thrive in the close company of humans. Members of the Corn Snake species complex are common sights in a wide range of situations, from the remotest corners of national forest to the crowded neighborhoods of Key West, where hardly a square foot of dry ground has not been claimed by some homeowner or business.

The character of the central Louisiana pine woods has been irrevocably altered, but the region remains steadfastly rural. The distinctive *P. slowinskii* bears the pigment signature of natural selection within these forests, yet retains enough of its ecological generalist ancestry to thrive in the farmland pastures and loblolly plantations that have supplanted the longleaf woods. The taxon's persistence may lead to its demise, however, in much the same way as the distinctive snake subspecies of south Florida are being subsumed by the encroachment of closely related phenotypes as the ecological heterogeneity of that area is homogenized by human tampering. Formerly recognizable forms such as the Florida Kingsnake, *Lampropeltis getula floridana*, Everglades Rat Snake, *P. obsoletus rossalleni*, and the Florida Key color variant of the Eastern Corn Snake have become less well defined as the sharp natural features that shaped them are dulled and equalized by development. Similarly, if Slowinski's Corn Snake maintains its presence, its individual beauty and integrity may be diminished as physiographic features creating gene flow barriers dissipate in the commercial pinelands of the western gulf coastal plain.

Description

Slowinski's Corn Snake is a member of the *Pantherophis guttatus* species complex (as defined by Burbrink 2002) outwardly differing from sister taxa by its distinctive coloration and inwardly differing by its DNA. Pattern elements are largely congruent with *P. guttatus* and *P. emoryi*. Typical specimens have 33–36 squarish dorsal blotches, with a series of smaller, more irregularly shaped ovoid blotches on the flanks aligned with the dorsal blotch interspaces. The head bears the distinctive spearpoint marking that distinguishes these snakes (Conant and Collins 1991). The venter is white, sometimes with pinkish overtones, and boldly patterned with black markings (plate 3.6).

Although *P. slowinskii* is similar in pattern to other members of the Corn Snake complex, the species differs in several respects. The dark borders around the dorsal blotches are consistently thinner on *P. slowinskii* than on *P. guttatus*, comprising less than half the length of a single scale (Burbrink 2002). On *P. guttatus*, the prominent black borders are a full scale in width. Ventral pattern is subtly distinctive. Rather than the bold, somewhat randomized black checkerboard pattern that is one of the hallmarks of *P. guttatus*, Slowinski's Corn Snake has smaller, slightly muted belly markings arranged in side-by-side pairs that form a vaguely defined double longitudinal line running head to tail on the ventral scutes.

Adults average 91 cm in length, but specimens exceeding 150 cm are common.

Ground color in adults is medium grayish brown. The dorsal and lateral blotches are a dark cherrywood brown with black borders. The contrast in hue between the blotches and ground color is weak; thus from a distance of several meters Slowinski's Corn Snake appears to be a unicolored brown serpent. The contrast between the blotches and ground color is variable, with some specimens so dark and muted that the blotches are hard to discern. Such dingy examples closely resemble the local dark color phase of the Prairie Kingsnake, *Lampropeltis calligaster*. Specimens in a minority subset have clearly seen markings that exhibit reddish hues; except for the distinctive ventral pattern, these are indistinguishable from dark examples of *P. guttatus* found further east. These snakes blend quite well with the pine needles that usually litter their substrate, and are very cryptic when crawling out in the forest.

Distribution Notes

Recent attention directed toward this species has refined our understanding of its distribution. Slowinski's Corn Snakes are restricted to the trans-Mississippi pine ridges and savannas of west central Louisiana and extreme eastern Texas. Older distribution maps portrayed a pattern very similar to that of the Louisiana Slimy Salamander, showing it to range into southeastern Arkansas (Conant and Collins 1991: map 167). The taxon was included on the list of Arkansan herpetofauna prior to the proposal that the population represented a discrete evolutionary entity, thus preceding the more focused definition of the form. The southern Arkansas snakes, resembling

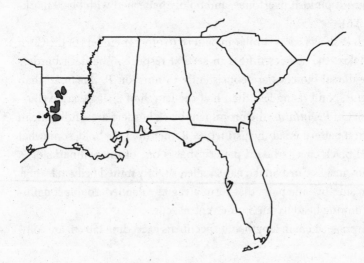

the dark Louisiana specimens, were therefore grouped together under the assumption that they represented a broad zone of intergradation of eastern and western subspecies as defined at the time. During his reanalysis of the entire complex, Burbrink (2002) questioned the identity of specimens from Arkansas and indicated that their inclusion as *P. slowinskii* remains an open issue. However, the proximity of these Arkansan rat snakes to *P. emoryi* lying to the west, and the apparent hiatus in documented collections separating these problematic populations from the defined *P. slowinskii* in central Louisiana, favors a tentative determination that the distribution of *P. slowinskii* is limited to Louisiana and Texas.

Habitat

Pantherophis slowinskii inhabits the rolling pine woodlands of north central Louisiana. This habitat begins in Vernon and Rapides parishes as the coastal flatlands of south Louisiana give way to a different terrain. Instead of flatwoods interspersed with bayous, sloughs, and swamps, the region inhabited by Slowinski's Corn Snake is delineated at its southern border by dry ridges intervened by savannas. This territory was forever altered by devastating clearcut logging at the beginning of the last century, and remains today a land of commercial forests. Slowinski's Corn Snake has adapted to these alterations well. The snake is abundant in monoculture loblolly tracts, as well as in cleared sites adjacent to such areas, particularly after vegetational succession has progressed to create grassy meadow habitat. *Pantherophis slowinskii* is particularly abundant at ecotones where shaded pine forest interfaces with open areas. Such situations are very common in north central Louisiana. The species is also fond of manmade refuges such as abandoned buildings and the ubiquitous rural trash dumps found along many secondary unpaved roads. In these refuse piles they are found under boards, corrugated metal, mattresses, and other objects that retain moisture and warmth.

In contrast, *P. slowinskii* is difficult to find in pristine areas, where it is more hyperdispersed. Here it makes use of stump holes. These snakes follow the deep crevices left after exposed roots have begun to disintegrate, using them as convenient avenues to escape winter temperatures. During the spring, a good way to find this species is to examine these structures during warm mornings when the snakes are likely to be basking. During the summer, the snakes become nocturnal and are one of the most common snakes found on blacktop roads in the region.

Ecology and Natural History

Slowinski's Corn Snake is ecologically more similar to its western relative, *P. emoryi*, than to the Eastern Corn Snake, *P. guttatus*. Unlike *P. guttatus*, *P. slowinskii* is not typically arboreal. Of dozens of specimens found in situ, all were on the ground and most were in open habitats where few trees were present. During a two-year trapping effort of snakes in north central Louisiana, traps in well-wooded habitat that *P. guttatus* would frequent did not produce specimens of *P. slowinskii*. Pine woods are generally poor environments for arboreal serpents, with their high, sparse canopies and limited branching and snags inherent in pine trees. This feature forces tree-dwelling snakes to remain partially visible to predators and prey and exposed to the direct heat of the sun. In hardwood and mixed pine-hardwood communities where *P. guttatus* is often found, the arboreal stratum of the landscape is festooned with good refugia such as rotting limbs, hollows, and forking branches, with the added security of dense leaf cover.

Gravid females have been found in June. Clutch size ranges from 12 to 14. Hatchlings begin appearing in August. For the first year they are colored much more brightly than adults, with most displaying some of the reddish hues typical of *P. guttatus* neonates. By the time they have reached 61 cm in length, Slowinski's Corn Snakes have taken on their distinguishing muted, darkened coloration.

Conservation

Slowinski's Corn Snake is part of a small subset of pine woods herpetofauna not in need of focused conservation measures. The common denominator among these taxa, which also includes the Pine Woods Treefrog, Oak Toad, Gulf Coast Box Turtle, and Brownchin Racer, is that they all commonly reside in or near human habitations or areas heavily impacted by human activity. Sampling in Louisiana indicates that Slowinski's Corn Snake is more abundant in disturbed situations such as logging clearcuts, rural trash dumps, or barnyards, than in pristine pine ridges and savannas. Disruptions in forests that create open fields, coupled with increased food sources of the type provided by agriculture, livestock, or human garbage, attracts rodents and supports higher densities of some snake species than in native habitat. The increased rodent population attracts and supports snakes such as *Pantherophis* and *Coluber*, as does artificial refugia in the form of lumber, woodpiles, and other ground debris. The distinction in habitat adaptability between specialists like the Apalachicola Lowlands Kingsnake and generalists like the present species is that the former taxon will not be found close to highly disturbed areas even when suitable cover exists, whereas Slowinski's Corn Snake is equally at home whether in a remote corner of national forest or in a rural backyard.

Summary

Slowinski's Corn Snake offers a unique glimpse into the processes of evolution, presenting an organism distinctly differentiated from its nearest relatives, yet still the same in most ways. Whether it warrants full species rank, subspecific designation, or is considered a geographic variation of the Eastern Corn Snake species depends on how one views the concept of the species, on which there is no consensus among biologists. What is certain is that this pine woods version of the Corn Snake is being shaped by its isolation in the trans-Mississippi sandhills and will continue to depart from related taxa if the isolation persists. Given enough time, measured in thousands of years, Slowinski's Corn Snake should evolve into a form quite divergent from rat snakes adapted to different environments; examining a present-day specimen is like viewing a child's scrapbook and wondering how the person in the photograph will look and act when adult—we may see hints of that future countenance in the immature face, but we know that only with time will the full form develop. However, one must consider the rapid pace of environmental change now taking place. The trans-Mississippi sandhills in the late 1800s looked very much then as they did 3,000 years earlier, but it is impossible to imagine that the rolling pine hills of central Louisiana will look anything the way they do today in 100 years, let alone a few thousand. Therefore, the possible consequences of continued specialization and speciation in Slowinski's Corn Snake may be merely an academic exercise in speculation.

4

Flatwoods Specialists

Flatwoods Salamander
Ambystoma cingulatum

The Flatwoods Salamander seems an unlikely center of controversy. It is small and shy, harmless in all respects to humans, and rarely seen by most people who live near it. How could such a nonentity in the consciousness of most people elicit arguments and divide communities into two fiercely battling armies? How could such a creature be the focus of a public meeting, attended by a multitude of people voicing sharp words of distrust and indignation against conservationists? The seemingly inappropriate role of the little Flatwoods Salamander as a flashpoint for controversy is due to its status as the only federally protected pine woods reptile or amphibian to occur extensively on private lands. A chronic issue that often erupts when an organism is considered for federal protection under the Endangered Species Act is whether the conservation needs of a species or habitat should restrict the activities of private landowners. This profoundly dividing issue has been raised in many pine woods communities through concern over the future survival of the Flatwoods Salamander.

On April 15, 1998, the U.S. Fish and Wildlife Service convened a public hearing concerning the listing of the Flatwoods Salamander as a Threatened Species as defined by the Endangered Species Act. Twenty-eight people in the audience spoke out on their concerns or opinions regarding this proposed action, only one in support of it. The factor that seemed to have been the primary motivator for most people to attend and express an opinion was the fear that activities such as timber cultivation and harvest, agriculture, and clearing would be hampered or prohibited if the salamander were given federal protection. This was a realistic expectation, as these activities, when unrestrained, are clear threats not only to this particular salamander but also to many components of the flatwoods ecosystem. The opposition correctly pointed out that no scientific study has directly linked such land utilization to the extinction or decline of any specific Flatwoods Salamander population. However, the deleterious consequences of these activities for the fundamental components of the flatwoods environment are indeed well studied, and it is logical to expect the Flatwoods Salamander to be affected very directly by these changes. An objective eye can see that a pine plantation and a natural flatwoods community are different environments.

Making bogus claims about Flatwoods Salamander abundance or resistance to habitat alteration polarizes stakeholders in the conservation issue and does nothing to bring opposing parties closer to satisfactory compromise. The problem of conserving the Flatwoods Salamander in a region where the economy and residents' livelihoods depend heavily on forest industry is vexing. The solution lies in finding common ground and reaching compromise, and implementing a set of actions that preserve the essentials of what each side holds most valuable. It also requires each individual or organization to concede in equal measure, so that both camps give "'til it hurts" and yet neither is defeated. Such an agreement is possible. There is no reason why a commercial silviculturist could not control competitive undergrowth primarily with fire, maintain extensive commercial stands while leaving any wetland sites intact and uncultivated, and harvest a good crop without destroying the upper soil layers in the process, but such an operation would certainly not be achieving maximum profitability.

Some of the hearing attendees who were fiercely opposed to federal listing were not arguing against protective measures for the salamander; conservation issues are never so black and white. Speakers from both sides of the issue made references to voluntary compromise agreements, or the formal Candidate Conservation Agreement (CCA) program administered by the U.S. Fish and Wildlife Service, as the most preferable way to protect the salamander. CCA's have been used productively in other conservation dilemmas involving land use practice, particularly when private acreage is key to success.

A CCA is an agreement between a landowner or management authority (in the case of public lands) and the U.S. Fish and Wildlife Service that outlines negotiated practices and management techniques that the Service deems sufficient to maintain the species within that tract. The CCA is less one-sided in perspective than a recovery plan mandated after a species is listed as Endangered or Threatened, and thus avoids the approach to land management that polarizes communities when federal listing of a local species is announced. A CCA can be developed prior to federal listing in an attempt to preclude the need for listing by securing the rare species and promoting its recovery before an eleventh-hour intervention is needed. Both government biologists and the landowner discuss and negotiate how conservation actions will be applied and which potentially detrimental practices will be modified, restricted, or abandoned. The landowner sits at the negotiating table as an equal partner with the government, and both parties work together to arrive at a solution. The Service plays its card of not approving a CCA it considers inadequate to protect the threatened flora or fauna, and the landowner plays an equal card, his or her freedom to decline any recommendation that might hamper valued activities and to walk away from discussions at any time.

The incentive for the Service to enact a CCA rather than take the road to listing is that placing a species on the list of Endangered Species is a long and arduous process. The path to listing is fraught with such protracted legal battles that even if the species is eventually listed, it may be so long in coming that crucial habitat and populations may dwindle beyond recovery. Indeed, some speakers at the hearing made provocative threats to sell their land to developers so any salamanders present would be extirpated before the regulations went into effect. Another incentive for the Fish and Wildlife Service to work hard to secure CCA's is that federal listing, even for the most obvious candidates, is far from assured and this tendency grows increasingly strong with each passing year. Shrinking natural resources and a growing human population are nurturing a growing opposition to the basic tenets and mechanisms of the Endangered Species Act, particularly in regard to the resource-rich and development-friendly southern pine woods. Increasingly, voluntary compromise agreements are becoming the only realistic means of conserving habitats.

From the perspective of a private landowner with a candidate Endangered Species living on the property, a CCA is vastly preferable to federal listing because the recovery steps that will be administered under a CCA will have been discussed and modified to the landowner's satisfaction. Under some types of CCA plans, should the species eventually be listed only the actions listed in the CCA will be required of the particular landowner, even if practices at other sites are more strict. CCA's impart a hopeful spirit into conservation biology, with two disparate parties working together to solve a problem, rather than entering into a pitched battle where a good outcome for the organism requires that the very entities holding land harboring it be beaten back and dominated, causing opposing positions to harden before the next issue inevitably arises somewhere else. Best of all, with a well-devised CCA, the imperiled species wins and its survival is maintained, at least within the jurisdiction of the plan.

Description

This species bears a resemblance to Mabee's Salamander, *Ambystoma mabeei*. Both salamanders exhibit fine silvery, white, or grayish reticulations over a darker ground color, but in *A. cingulatum* the pattern is more uniformly distributed such that no faint barring or blotching is usually apparent, as is typically seen in *A. mabeei*. In addition, in both reticulated markings as well as ground color, Flatwoods Salamanders tend more strongly toward hues of gray, silver, and black, while Mabee's Salamanders trend more toward tones of brown (plate 4.1).

The silvery pattern makes it appear as if a finely woven fish net had been cast over the salamander. Upon close examination, the light lines are seen to be composed of silvery flecks. These markings extend over all aspects of the salamander's dorsal surface, including tail, sides, and limbs. The ventral surface of adults is medium to dark gray, sometimes faintly marked with lighter areas but in no definable pattern.

Adults range from 10.2 to 15.2 cm in total length.

Within their characteristic habitat, *A. cingulatum* is the only salamander whose larvae are so boldly striped. A light tan or yellowish middorsal stripe is flanked on both sides by a dark gray or black band, each of which is bordered by another light stripe, followed by a black stripe along the lower flanks. This arrangement of 7 bold stripes makes identification of *A. cingulatum* larvae in coastal flatwoods straightforward.

Distribution Notes

This is another of those species for which a sharp distinction between historical and current distribution must be emphasized, because along with the Dusky Gopher Frog, Louisiana Pine Snake, and Southern Hognose Snake, the evaporation of huge areas where the species was once known to occur has been truly alarming. At one time, the species was known to inhabit numerous flatwoods sites within the southern pine belt, extending from southeastern South Carolina through southern Georgia, northern Florida and extreme southern Alabama (Conant and Collins 1991). Suitable habitat was an historically abundant feature throughout this region, and old locality records are sprinkled extensively within it. The Florida distribution was the largest by state, including most of the panhandle and extending south at least as far as Marion County (Palis 1997b). Today, however, Flatwoods Salamander populations are reduced to an easily countable number and often spaced across hundreds of miles of intervening uninhabited territory.

The current distribution of the species is well documented, largely because historical sites and potential new sites were surveyed during the years leading up to federal listing, while the species was still a Candidate for protection. At the time of its listing as a Threatened Species in 1999, 51 extant populations were known, the bulk of them, 36, situated in Florida. Of the remainder, 11 were in Georgia and 4 in South Carolina (LaClaire 1999). The U.S. Fish and Wildlife Service identified 110 historical records since the species' description through 1990, but when most of these sites were revisited post-1990, only 12 could be confirmed as still extant. None of the Alabama locations known to once harbor the salamander were still supporting populations during recent surveys, and no new sites have been found, so the Flatwoods Salamander appears to have been extirpated in the western portion of its range.

The most contiguous areas inhabited by this salamander are in the Florida panhandle, where more than 21 percent of the known populations occur. These low-lying rural regions still contain extensive flatwoods, particularly on the extensive protected public lands. The two epicenters of modern-day

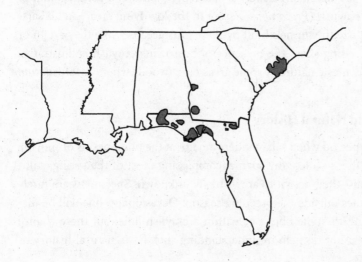

occurrence are in Okaloosa and Liberty counties, where Eglin Air Force Base and Apalachicola National Forest are respectively situated. These two areas are especially significant because, unlike many populations elsewhere, the Eglin and Apalachicola populations are associated with numerous breeding sites and thus are probably more robust and resistant to local extinctions than most other populations dependent on single breeding pools.

Habitat

This is an obligate inhabitant of pristine flatwoods, as its reproductive biology dictates the consistent recurrence of winter rain-filled depressions exhibiting peculiar flatwoods characteristics. Beyond the general conditions afforded by good flatwoods habitat, potential A. cingulatum sites are suggested when longleaf pine dominance gives way to the more mesophilic slash pine in depressions, indicating the probable location of ephemeral winter pools. Breeding ponds tend to be long-standing depressions that have held water off and on for many seasons, as indicated by the development of more moisture-dependent vegetation than is typical of the surrounding area. Trees such as blackgum, *Nyssa sylvatica*, and cypress, *Taxodium ascendens*, when growing in aggregations, indicate the likely presence of winter ephemeral pools. Periodic fires sufficient to maintain an open understory primarily of wiregrass is critical, as larval salamanders require this as cover and refuge from predators. This salamander does not appear capable of persisting in otherwise suitable habitat where fire-suppression has allowed woody undergrowth to preclude open, grassy breeding pools (LaClaire 1999).

Another predictor of Flatwoods Salamander sites is a congregation of burrowing crayfish (*Procambarus* spp.) in the low-lying areas. Adult salamanders utilize the tunnels built by these crustaceans while they are in or near the breeding sites. The presence of burrowing crayfish is illustrative of the highly mesic nature of some areas of flatwoods where A. *cingulatum* occurs.

Ecology and Natural History

During the period when salamanders are not at the pond basins or moving to or from them—a lengthy span encompassing most of the spring, summer, and fall—their activities are virtually unknown. Specimens are rarely found at times outside the breeding season. Occasionally, one will be discovered beneath some object or within a crayfish hole, but these chance encounters are of no help in understanding such basic natural history as preferred prey and means of procurement, home range size, microenvironmental preferences, and daily activity patterns. They are presumed to scatter from the breeding ponds—both adults and metamorphs—to move into adjacent drier uplands. Some feeding and movement must occur during the nonbreeding season, but periods of aestivation during prolonged hot, dry weather may also account for a significant amount of time.

Considering how important the winter ponds are to the Flatwoods Salamander, it is surprising how little time an individual spends at these sites. Migration to the low-lying areas that become ephemeral pools is triggered by heavy rains, usually occurring between October and December (Palis 1997a). At any specific pond, the migration events last only a week or two. Adults arrive at the pond sites as the depressions are filling, but before the water level inundates the vegetated margins, as these must not be submerged when *A. cingulatum* deposits its eggs.

Eggs are laid in and around clumps of wiregrass, under leaf litter, or within crayfish burrows, at locations where they will be underwater once the pond fills completely. Most individuals do not linger at the pond after breeding and ovipositing, but quickly return to upland retreats under the same rainy conditions that first called them to the site. Although eggs are laid on dry land, hatching occurs only if accumulating rainfall submerges them. As unreliable as surface water accumulation is in the southern pine woods, Flatwoods Salamanders are unable to breed in some years, and when rainy weather is sufficient to stir migration, it is sometimes still inadequate to fill the basin to the level that inundates the eggs, resulting in no recruitment from a breeding pond. Conversely, the requirement that the vegetated margins of the ponds remain dry during the late fall and early winter, so that courtship, breeding, and oviposition can occur on land, means that during years with unusually abundant rainfall, ponds may not dry during the summer, resulting in early filling when the salamanders begin their breeding migrations. If the pond is already filled when the adults arrive, there will be no suitable egg-laying sites on dry land, resulting in an unsuccessful reproductive season.

The reproductive biology of this species is illustrative of the very restrictive parameters within which amphibians endemic to pine woods replenish their populations. To the uninformed, so long as bodies of water appear in the forest during wet winters, the site would seem ideal for Flatwoods Salamanders. However, this species is extremely particular in the characteristics that must be present for breeding to occur. Temperature fluctuations and precisely timed periods of dry versus rainy weather are additional factors that are crucial to reproductive success but over which forest managers have no control. Even if all these elements work in concert, if migrating adults are forced to traverse landscape made inhospitable by intense silviculture practices or poor land management, reproduction may be prevented (Means et al. 1996).

Conservation

As a species that utilizes both terrestrial refugia in xeric habitat as well as transitory aquatic situations, the Flatwoods Salamander is vulnerable to impact by a wide variety of land management practices and misuse. Its inability to flee quickly from approaching danger and disturbance renders it especially sensitive to small-scale disruption of its immediate environment. Amphibians are at high risk for poisoning by environmental contaminants via direct absorption through their permeable skin and their proclivity for living in close contact with moist substrate or in surface water catchments. Consequently, in addition to the rampant destruction of coastal pine wood habitat that endangers all flatwoods plants and animals, *A. cingulatum* populations face an additional array of problems that attack its specific vulnerabilities.

To transform poor-draining flatwoods into land suitable for real estate development or commercial silviculture, trenches or ditches are often dug to produce drier conditions. In addition to destroying mesic microrefugia critical to salamander survival during dry summer months, such features also disrupt or completely obliterate hydroperiods in rainfall catchments, which is exactly what they are designed to do. Such practices make suitable winter breeding sites even scarcer than they naturally are. Populations that do persist are sometimes constricted within a time period that is too short for larval development to reach completion (Means et al. 1996; Palis 1997b).

Where Flatwoods Salamander populations have managed to coexist with commercial silviculture, two common features of loblolly monoculture pose a threat. Because breeding ponds are usually no more than low places on the forest floor where winter rains collect, the water that fills them comes from surface runoff over a wide area; in essence, dirty rinse water if contaminants are present. Chemical residues left by fertilization of surrounding stands often make their way into these ponds; depending on the attributes of the local terrain, they may reach high concentrations. Once within the pond, these excessive nutrients can cause an algal bloom in the normally clear and well-oxygenated water, suffocating much of the natural fauna and making it unfit for larval amphibians (Palis 1997b). In a similar fashion, herbicides used in pine plantations to suppress competitive plant growth, as well as pesticides, often make their way into breeding ponds.

Salamander survival during the summer months is dependent on the ready availability of mesic microenvironments. The root networks of herbaceous vegetation in regularly burned forests produce avenues to moist environments in the deeper layers of substrate that are otherwise inaccessible to such poor diggers as small salamanders. When dense loblolly or slash pine plantations reach a size where crown closure begins to take place, the young trees are still too small to withstand a prescribed burn. Thus follows an extended period of up to 10 years of fire suppression, even on lands where prescribed burning is utilized. The closed canopy, fire-suppressed stands quickly shade out understory vegetation, resulting in a forest floor without any appreciable flora and carpeted only with a thick layer of pine straw. With young pine trees as the only vegetation, dead tree falls and resulting rotting logs are virtually absent. Such forests, with their immaculate, coverless ground devoid of mesic refugia, are hostile environments to Flatwoods Salamanders, particularly during the hot, dry months of summer (Means et al. 1996).

Another form of shelter for Flatwoods Salamanders are the tunnels of fossorial crayfish, common along the interface of wetland and drier flatwoods environments. Crayfish are a valuable economic resource to many landowners, and they are often harvested. Naturally occurring populations of crayfish are easily exhausted by overcollection. Little consideration is given to the impact on local amphibians when crayfish are harvested for the seafood market. Where local colonies are eliminated, an important summertime refuge for pineland salamanders is lost.

A frequently unnoticed aspect of pine woods conservation is the importance of maintaining the integrity of its peculiar substrate. Many of the small and secretive reptiles and amphibians, as well as invertebrates, depend on favorable conditions for burrowing and moist refuges afforded by delicate pine woods ground and groundcover vegetation. Species like *A. cingulatum* are sedentary and unable to quickly flee when their immediate surroundings are disrupted; thus any human activity that disturbs the upper few centimeters of soil is a threat to these creatures. The most destructive practice applied to these crucial upper layers of ground is roller chopping.

This procedure, routine on many pine plantations, consists of cutting and churning of the soil with machinery following a timber harvest, in order to eliminate large stumps and roots and loosen the soil to improve conditions for replanting. Roller chopping is employed on a large scale, encompassing many acres at a single site, and obliterates the natural subterranean micro-habitats on which salamanders depend. On tracts where timber is harvested but not roller chopped, the substrate is less brutalized, and the remaining stumps provide retreats for many types of fauna.

The construction of roads, even those consisting of no more than tire tracks on sand, can have a damaging effect on salamanders if they traverse critical habitat. Trails worn down by all-terrain vehicles can also cause mortalities, either directly or by reducing ground cover and compacting the soil so tightly that burrows cannot be made and those already present are crushed. Even firebreaks, which ironically are part of the beneficial practice of prescribed burning, must be carefully situated so the surface and subsurface microenvironments are not damaged in areas inhabited by salamanders. Dry slopes adjacent to ephemeral wetlands are an example of the type of area that should be carefully assessed before any sort of road, trail, or clearing is established.

The common practice of prescribing burns during the winter months, favored by foresters because of the cooler and wetter conditions that make controlling burns easier, may pose a significant danger to salamanders lying dormant near the surface. During cold days, salamanders are vulnerable to injury from fire due to torpor and because the moister conditions enable them to be closer to the surface. During the summer months, when fire is more effective in controlling vegetative overgrowth, salamanders are residing deeper underground to escape desiccation and are better able to escape even deeper into the ground as fire passes overhead.

In 1982, the U.S. Fish and Wildlife Service designated A. cingulatum a species possibly in need of federal legislation to protect wild populations but, because of a perceived lack of scientific evidence, one for which no immediate action would be taken. This limbo status remained in effect until 1996, when the Category 2 conservation status was discontinued. With the elimination of most Category 2 species, the Flatwoods Salamander and a host of other rare organisms were set aside from official concern in federal conservation strategies. In 1997, a petition to list the Flatwoods Salamander as an Endangered or Threatened Species was submitted to the U.S. Fish and Wildlife Service by conservationists, but the petition was judged to provide insufficient evidence to warrant federal protection and the petition was denied. Two years later, the petitioners sued the federal government for violation of the Endangered Species Act, an action intended to force the issue into a court for a judge's decision. The result of this suit was a reconsideration of the petition, and subsequently a finding that the listing might be warranted. This led to a period of public comment, as a required part of the process at this stage, including two public hearings and the receipt of nearly 200 written or oral responses from a wide range of parties. The U.S. Fish and Wildlife Service weighed the accumulated evidence and commentary, and determined that federal action was needed. On May 3, 1999, the Flatwoods Salamander was designated a Threatened Species under provisions of the Endangered Species Act (LaClaire 1999).

Summary

If conservation of the southern pine woods can be effective in saving the small, secretive, and seemingly unimportant species dependent on them, it will first be apparent with the Flatwoods Salamander. Conservation efforts for this species are better organized and more advanced than for any other endemic herpetofauna. The conservation of more generally appealing and thus more marketable taxa as the Red-cockaded Woodpecker and Gopher Tortoise is promoted under a different set of parameters than such easily dismissed organisms as salamanders. No other reptile or amphibian restricted to pine woods has been favored with more attention. It receives both state and federal protection. A sizable portion of its remaining populations are living on federal lands that are well managed pineland habitat, places such as Eglin Air Force Base and Apalachicola National Forest, where prescribed burning and respect for humble-looking ephemeral wetlands are part of the standard operating procedure. An active consortium of dedicated biologists and habitat managers is composing a recovery team that devises and implements aggressive measures to protect key sites. Though lacking much fanfare, a daring experiment is being conducted on the cutting edge of amphibian conservation. The hypothesis being tested is that an obscure and unpopular species can be guaranteed a secure place among the remarkable fauna of the southern pine woods. If this experiment fails, it will be an ominous sign for every species highlighted in the pages of this book.

It is clear that the Flatwoods Salamander could benefit from a network of rangewide CCA's as a means to ensure its survival. One has only to consider the situation on Eglin Air Force Base in Okaloosa County, Florida.

There, military operations and training maneuvers take place unabated, yet the site is one of the largest and finest pine woods refuges anywhere. Eglin AFB is a superb example of human priorities and nature coexisting as long as conservation and the preservation of wild areas is a sincere value that guides land use and management decisions. The Flatwoods Salamander has not everywhere benefited from the spirit in force at Eglin Air Force Base, but time has not run out for the species.

Mabee's Salamander
Ambystoma mabeei

The southern pine woods are a habitat in decline, at a speed so rapid that transformations from pristine habitat to biologically decimated development occupy the brief span of one person's memory. I have personally observed the annihilation of savannas in Louisiana and Mississippi. The first and most disturbing experience of this type began with a two-week camping trip my brother and I took in 1976. We set out from Philadelphia in my car, bound for the herpetological treasure land of Jasper County, South Carolina, where my boyhood hero Carl Kauffeld, Staten Island Zoo's famous Curator of Reptiles, had enjoyed so many of his storied snake hunting adventures. This legendary land beckoned me with the promise of finding and holding in my hands the gems of the Carolina pinelands: diamondbacks, pine snakes, and corn snakes.

After a night's stop in Croatan National Forest in North Carolina, we made our way down Highway 1 through the coastal plain of North and South Carolina, stopping midway along the coast of South Carolina in an area of scrubby, wet flatwoods. Neither national forest nor private preserve, the environment was nevertheless in a pristine natural condition because of its rural and off-the-beaten-path locale. Although somebody must have owned the land, that person's claim and influence was not apparent, except for the blackened, litterless ground and charred pine trunks that might have been caused by a fire deliberately and wisely set.

We set up camp for the night in a low depression that, although a little wet, afforded the advantage of not being scorched by the recent fire; thus we could move about without becoming streaked with charcoal from brushing against burnt twigs and shrubs. Our tents and sleeping mats insulated us from the soggy ground and allowed us to enjoy and be comforted by our comparatively lush surroundings in the overall bare and blackened forest. We set about gathering fuel for our campfire. I had picked up perhaps two or three fragments of wood when beneath the next one, I exposed a beautiful Mabee's Salamander.

It was an exceptionally beautiful specimen I admired, sitting in my hand, and I called my brother over to see. This one had considerable silvery markings that faintly suggested a netted pattern. It was a very large male, with a stocky build and the disproportionately small head that is a hallmark of the species. After taking time to fully appreciate what I considered a once in a lifetime encounter, I gently returned the salamander to the slight depression in the mud and lightly replaced his pine limb cover.

Returning to our chore, we resumed gathering wood, but within a few minutes my brother called out that he too had found a Mabee's Salamander, pre-empting by seconds my exclamation upon discovering another. Abandoning wood gathering for good now, and turning to a full-hearted herpetological quest, we began an earnest search for salamanders, and within 20 minutes we were rewarded with four more. Here in this out-of-the-way spot in coastal South Carolina, we had stumbled on a thriving population of an infrequently seen amphibian. Its habitat still pristine by default, it had simply been fortunate to live on a site that had attracted no human interest.

Almost 20 years to the day after that memorable afternoon, I found myself once more in the pine flatwoods of South Carolina, on an excursion to find and photograph the Pine Woods Snake, *Rhadinaea flavilata*. I decided to return to the salamander site, hoping to see examples of *A. mabeei* as well. As I turned off the vaguely familiar blacktop and onto the gravel road that led to the site, I noticed a bright sky on the horizon through the trees, indicating that a clearing lay ahead. Leaving the forested stretch, I drove into a large open area carpeted with mowed grass and dominated in the center by a metal-sided utility structure that housed some sort of business or industry. Surrounded by a chain-link fence was a graveled lot with several parked vehicles. The building itself sat almost precisely where my brother and I had found the priceless biological treasure two decades before. The basis for the pine wetland that had once existed was now revealed, because the denuded treeless landscape exposed a slight depression in the land, where the building now sat, which would have attracted rainwater and retained it after the surrounding ground had dried. There was no need to reconnoiter the vicinity for salamanders, I realized, for it was obvious the population was gone. I stood at the side of the road, looking off to the horizon, back into my past, remembering a time for the pine flatwoods of this area that would not return, and then got in my car and drove away.

Description

Adult Mabee's Salamanders are 8–10 cm in total length, making them the smallest member of their family in eastern North America. In overall gross morphology they resemble the Mole Salamander, *A. talpoideum*, except for their relatively longer toes and, notably, an unusually small head for an ambystomatid (Conant and Collins 1991), instead of the large, broad head of the Mole Salamander. This species also bears a superficial resemblance to the Marbled Salamander, *A. opacum*, but differs by its broader trunk and tail, and again by its notably small head.

Base color ranges from deep brown to black, with lighter flecking. Both the ground color and flecking become lighter laterally, and the flecks become much more prominent and conspicuous progressively down each side. The flecks gradually change from grayish white to dull tan as they approach the dorsum, eventually becoming obscured by dark pigment. The virtually unspotted venter is light brown, and the gular region is lighter still, being a pale fleshy hue (plate 4.2).

Hardy and Olman (1974) gave a detailed description of the larvae. The important points of their account are as follows. A distinct light lateral stripe extends from the gills to the tip of the tail, frequently broken by darker cross bands. Above this stripe is a dark dorsolateral stripe containing irregular lighter spots. A light ventrolateral stripe is contained within 2 indistinct, rough-edged dark stripes. Both the venter and the cloacal regions are unpigmented, although tiny specks of pigment may be discernible near the midline. A dark stripe extends from the gills to the nostrils—postorbitally rough-edged, diffuse, and obscure—while preorbitally expanded and broad. The upper edge of the jaw is marked with small spots or dashes.

Distribution Notes

The species is restricted to the coastal plain of North and South Carolina and a small portion of southeastern Virginia (Conant and Collins 1991). The type locality of Dunn, North Carolina, lies at the approximate center of the north-south axis of the range, and very close to the western limit. At this midpoint in the range, the Fall Line is almost reached, but as the range goes progressively north and south, some authorities (Conant and Collins 1991)

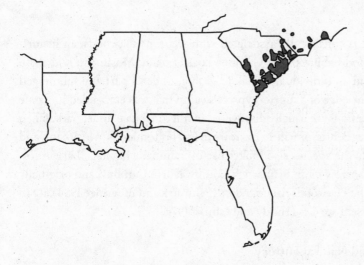

show it receding farther from this biotic boundary zone. Examination of vegetation and topographic maps reveals no reason that the species would not occur all the way to the Fall Line throughout its range. This salamander reaches its northern limit in roughly a dozen counties in southeastern Virginia. In addition, one specimen was found considerably farther west of the known range in South Carolina (Raymond Semlitsch, pers. comm.).

It is uncertain whether the species occurs in Georgia. Some authors have included Georgia in the distribution (Dunn 1940; Bishop 1943). Conant's (1958) range map shows a question mark in Georgia, indicating he recognized both the tenuous basis of this contention and the fact that no geomorphic or biotic condition seems to exist that would exclude Mabee's Salamander from extending its range slightly into Georgia. Those authors who do include Georgia in the species' distribution limit its occurrence to Liberty County. The only solid evidence for the inclusion of Georgia is traceable to the early works of Edward Drinker Cope. Cope sent two specimens to the Philadelphia Academy of Natural Sciences that he had acquired from John LeConte of Riceboro, Liberty County, Georgia, and assumed they had been collected there. However, LeConte sent many other specimens from his home base of Liberty County that had been collected elsewhere (Neill 1957). Since the only basis for including Georgia is at least doubtful, and no substantiation of the Liberty County records has come to light in 100 years, it seems prudent to do what Conant and Collins (1991) subsequently did, and delete Mabee's Salamander from the list of Georgia's herpetofauna.

Habitat

The habitat is consistently associated with wet pine flatwoods, an historically common habitat of the Carolina coastal plain. Mosimann and Rabb
(1948) found an individual in classic setting, under a partially submerged
stump on the edge of a tupelo-cypress swamp that was embedded in a pine
forest. Specimens are usually discovered in microhabitats under bark, fallen
logs, or other debris, or within burrows in low bottomlands (Mosimann and
Rabb 1948). The species also makes use of manmade habitats. Larvae collected in a borrow pit in Brunswick County, North Carolina, and originally
thought to be Flatwoods Salamanders (Schwartz and Etheridge 1954) are in
fact the present species (Hardy and Olman 1974).

Ecology and Natural History

The reproductive strategy is notable for the very early timing of migration
and breeding (Conant and Collins 1991). This strategy affords a salamander the advantage of leaving eggs in ephemeral pools immediately after the
pools fill, thereby increasing survivorship by providing more food and spatial resources because of lack of interspecific competition and a lower level
of large insect predators. It also gives the larvae more time to develop and
grow, so they can achieve maximum size and higher metamorph survivorship resulting from predator gape limitations, wider prey size suitability,
and reduced vulnerability to desiccation. Most observations of breeding migrations have been made in North Carolina, toward the northern limits of
the distribution, from late January through March, depending on the timing
of periods of heavy, warm rain showers. Populations farther south appear
to breed earlier in the year, as indicated by collections of migrating adults
in southeastern South Carolina during November and December (Russell
et al. 1998).

Adults migrate to breeding pools in autumn and mate and deposit eggs
before the depressions actually fill. The mass movement of 76 individuals was
observed on a highway near a known breeding pond during a warm, rainy
night in late March in North Carolina (Hardy 1969). Gravid females have
been found in dry pond bottoms in late autumn (Hardy 1969). Geographic
variation in this life history trait may occur, because breeding has been observed from late fall to early spring. Hardy (1969) interpreted a bimodal
size distribution of larvae in several breeding pools he studied as indicative
of two distinct breeding events that season. All larvae-occupied ponds ex-

amined by Hardy (1969) were ephemeral, with two of the five ponds drying before any larvae could transform, leading him to conclude that pond drying was a major factor controlling emergence success and population size in this species (plate 4.3).

The eggs may be laid singly or in chains of 2–6, and the 8.5 mm hatchlings emerge in 9–14 days. Larval transformation size ranges from 50 to 63 mm (Bishop 1943; Jobson 1940). An interesting observation centered on the discovery of 91 newly transformed juveniles under debris in an overgrown lot, with the closest ponds more than 0.8 km away (Hardy 1969). This suggests that the metamorphs migrate en masse from the breeding pools, as has been documented for Spotted Salamanders, *A. maculatum* (Hardy 1952).

Breeding pools may harbor adult and larval Tiger Salamanders, Lesser Sirens (*Siren intermedia*), and Red-spotted Newts (*Notophthalmus viridescens*) (Schwartz and Etheridge 1954). Hardy (1969) reported larvae in the stomachs of Tiger Salamander larvae, and stated that sirens would feed on larvae in the laboratory. It is well known that adult newts prey on salamander ova, and may exert considerable pressure on ambystomatid populations where the two genera are sympatric (Morin 1983). The rapid growth rate of larval ambystomatids, coupled with their relatively large minimum and maximum threshold sizes, means that newts are probably significant predators of eggs only.

Of all *Ambystoma* he examined, Hutchinson (1961) found that Mabee's Salamander had the highest critical thermal maximum, a surprising 38 C. This is 8 degrees higher than any microhabitat field temperatures taken! Mabee's Salamanders spend most of their time in mesic oases embedded within much drier landscapes, and Hardy (1969) found the species prevalent in hot, xeric situations in late spring and summer. Thus Mabee's Salamander appears to exhibit an adaptation to a pine woods habitat that, at first consideration, would appear unsuitable for such a delicate creature.

Conservation

The species is considered uncommon (Wright 1935a; Hardy and Anderson 1970). Many secretive organisms have been erroneously tagged "rare" before their habits became better known. However, only 12 years after its description, Jobson (1940) noted that because of the draining of swampy areas, Mabee's Salamander was no longer found in places where it had previously occurred. As with other components of the herpetofauna of the once great southern pine belt, this amphibian has had large parts of its range subjected to lumbering and clearing. Although no evidence exists of a decline in numbers or range as a result of these activities, it's likely that an organism as immobile and environmentally sensitive as a salamander would be even more vulnerable to the same practices that have virtually exterminated the tough old Gopher Tortoise from the Carolinas.

Summary

Described in 1928, Mabee's Salamander remains a poorly known animal, and is an obscurity among an otherwise well-studied genus. The reason for the void in knowledge of this species is not clear. It occurs in an accessible and otherwise well-studied area of the United States. It may simply be an uncommon species, living in low densities, although it should be remembered that until the work of Raymond Semlitsch and Whit Gibbons at the Savannah River Ecology Lab, the Mole Salamander was also considered uncommon, but we now know that it is abundant but uncommonly seen. In 1970, Hardy and Anderson said that Mabee's Salamander was rare in collections and pointed out that the geographic range is poorly defined, and the courtship had not been described, observations that remain true today. Mabee's Salamander is an enigma for future biologists to bring into clearer perspective, but with the Carolina flatwoods in such disrepair, the opportunities to study its natural history are not infinite.

Gulf Coast Box Turtle

Terrapene carolina major

This most distinctive of U.S. native box turtles sits in relative obscurity in the minds of many biologists. Few appreciate the suite of unique morphological characters that make this subspecies so recognizable. Fewer still are aware that it is so closely associated with a piney habitat. Of all pineland reptiles, the Gulf Coast Box Turtle is the one most likely to be left off the list of endemics by herpetologist compilers.

One reason for the low profile of *T. c. major* as a taxon of significance is what Conant and Collins (1991: 53) termed "the influence of *major*," referring to the broad areas of intergradation between this and other regional forms of *T. carolina*. The influence of *major* is indeed widespread and pervasive. Along a lengthy stretch of the Gulf coast, box turtles are not of the usual inland variety. In general, they tend to be larger than *T. c. bauri* in peninsular Florida or adjacent populations of *T. c. carolina* and *T. c. tringuis* found only a short distance from the coastal region, and display more muted carapace markings than the first two taxa. These populations are often assigned to *T. c. major*, but hardly warrant subspecific ranking given their polymorphism and the similarity of some variants to the other three subspecies in the southeast. As a consequence of this wide zone of intergradation, *T. c. major* is sometimes considered a weakly defined race.

Another source of misconception regarding the defining characters of the Gulf Coast Box Turtle is the ontogenetic progression of color and pattern change. Hatchlings are hardly distinguishable from other subspecies and do not begin to display the remarkable subspecific diagnostics until early adulthood. Thus, biologists considering box turtles collected along a large swath of the southeastern Gulf coast, and including subadults in their samples, will be led to the impression that the Gulf Coast Box Turtle is a highly variable and poorly defined subspecies of *T. carolina*. True *T. c. major* from the core of their range are divergent from conspecifics in uniform and striking ways, and immediately identifiable as a distinct entity from a specific environment.

Description

This race is the largest of all the box turtles. Males may exceed 23 cm in carapace length. A hallmark of the subspecies is the upwardly flared rear margin of the carapace. Coloration is distinctively somber. Adult males have a dark slate gray carapace that is unmarked or nearly so. Females usually retain traces of yellow dots or streaks that are even more prominent in juveniles. Singularly seen in this race is a tendency toward white blotching on the head and gular region of adults, sometimes coalescing to produce mostly white-headed specimens. The combination of large size, dark unpatterned shell, and white head makes adult Gulf Coast Box Turtles the most distinctive-looking member of a genus otherwise notable for phenotypic conformity. The unique characters of this specialist of the pine flatwoods and dry coastal sandylands of the extreme Gulf coast are not well appreciated because of the restricted distribution of populations free of genetic influence from other subspecies (plates 4.4, 4.5).

Distribution Notes

Though the area of occurrence of *T. c. major* is often described as Gulf coastal plain from Florida to Texas, this is deceptive because it delineates

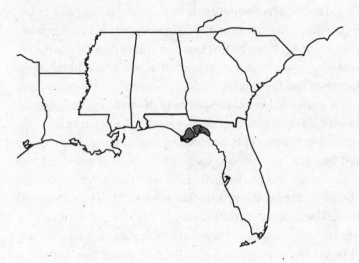

the range with too coarse a brush. True *T. c. major*, showing all the behavioral, physiological, and morphological characteristics that make it a highly distinctive form, occur only in the Apalachicola River drainage basin and adjacent coastal and offshore habitat. This area is entirely encompassed within Wakulla, Liberty, Franklin, Gulf, and Calhoun counties. Elsewhere along the Gulf coast, and in habitats other than pine flatwoods, box turtles exhibiting some, but not all of the characters of *T. c. major* can be found.

This taxon is believed to be most closely aligned with the Florida endemic *T. c. bauri*, on the basis of hind foot and phalangeal morphologies (Minx 1992). Minx further stated that both of these subspecies stand distinctly apart from the other eastern conspecific, *T. c. triunguis*.

Habitat

The small area inhabited by pure *T. c. major* is a region contrasted by xeric coastal strand and barrier island dunes juxtaposed against soggy flatwoods prone to frequent flooding interspersed with marshland and estuaries. The Gulf Coast Box Turtle exploits both types of habitat. Immediately beyond the beaches in the undeveloped stretches of the Apalachicola coastline there is often a transitional zone that gives way to the flatwoods and marshes that dominate the interior. These transitional areas are primarily scrub oak thickets covering old dunes, with sparse sand pines. The dune caps are sparsely vegetated, but their steep slopes lead down ravines with a nearly impenetrable mesh of woody growth at the bottom. Gulf Coast Box Turtles are abundant in these situations if development has not encroached too closely. Unfortunately these dunes, with their gulf vistas and location set back just far enough from the surf to be fairly well protected from storm surge, are prime targets for real estate development.

The second, more ubiquitous habitat association for *T. c. major* is the wetter pine flatwoods farther inland. In addition to flatwoods proper, many of the ecotones between flatwoods and marshland provide good habitat for box turtles. The tremendous expanse of protected acreage of the Apalachicola National Forest and other natural habitat preserves in the area harbor Gulf Coast Box Turtles in great abundance (plate 4.6).

Ecology and Natural History

The mild climate of Florida's Big Bend region allows *T. c. major* to be reproductively active over a longer time span than conspecifics. Mating commences in March and continues through April (Jackson 1991). Clutches are produced at 3- or 4-week intervals, well into August. Winter egg diapause is not known for *T. c. major*, suggesting that hatchlings emerge as late as November. In most ways, courtship mirrors the generic pattern, with the possible exception of more olfactory focused behavior. Ernst (1981) observed males go through a phase of courtship where their interest was primarily on the cloacal and shell bridge areas of females. The males were interpreted to be searching for sexual scent cues. *Terrapene c. major* often courts and copulates in shallow water (Ernst 1981)—not surprising given the low, flat, coastal landscape in which this turtle lives.

As many as 5 clutches have been recorded from a single female during the extended reproductive season (Tucker et al. 1978), but no direct observations of nesting sites have been made. Because of the distinctive habitat occupied, suitable nesting sites are likely to be more of a limiting resource for *T. c. major* than for other eastern *Terrapene*.

The Gulf Coast Box Turtle is an enigma among pine woods herpetofauna in that it appears to be quite sensitive to dehydration, more so than closely related forms in other habitats. In a study of intrageneric evaporative water loss rates, Morgareidge and Hammel (1975) found that *T. c. major* had a higher rate of dehydration than *T. c. carolina*, *T. c. triunguis*, or *T. ornata*. Especially during the spring, when pools and ditches are flooded, I have frequently observed *T. c. major* resting partially submerged along the shallow margins of such pools.

It is tempting to explain the dark somber coloration of this subspecies as an adaptive feature. Fire has been an ecological force in flatwoods from the earliest origin of this habitat type. Even a cursory stroll through natural or properly managed pine woods of the region will show the ground to be littered with charred objects, stumps still anchored in the earth while limbs and fallen trees are shattered into dozens of black chunks across the forest floor. In such situations, the mind filters out these ubiquitous features, and a motionless box turtle colored in sooty hues is easily overlooked.

Conservation

Fortunately for the Gulf Coast Box Turtle, its habitat preferences concentrate most populations in areas of limited utility for human use. Poor drainage characterizes much of the flatwoods of the Apalachicola basin; they are comparatively devoid of development and remain rural. Furthermore, a large part of the area of distribution of true *T. c. major* phenotype is protected by Apalachicola National Forest and a network of surrounding protected areas. A drive along Florida's Gulf coast, from Port St. Joe to Cedar Key, will amaze anyone who has witnessed the steady degradation of southern pine-dominated habitats with its uniquely undisturbed character.

The Gulf Coast Box Turtle is the only pine woods reptile or amphibian for which the primary factor threatening its survival has been decisively addressed. This is a species whose security in the wild has actually improved over the past decade; no other endemic of this habitat has been so fortunate. Refreshingly, habitat destruction does not figure prominently in a review of the conservation status of this taxon.

Although the subspecies is abundant and conspicuous throughout its core range, two processes do pose a threat. The first is the high mortality caused by roadkill. During certain seasons, or at any time immediately following warm rain showers, box turtle activity increases dramatically, and one of the consequences is an abundance of turtles on roads. A morning drive down the rural roadways in this part of Florida will often reveal a dozen or more *T. c. major* crossing or basking on the asphalt. A large proportion of these turtles will be killed or injured by callous or incompetent drivers. In the early spring, most of these specimens are males searching for mates. Later in the summer, the bias tilts toward females, either gravid or returning to feeding grounds after nesting. As with *Terrapene* elsewhere, few juveniles are encountered, and finding a juvenile on the road, or for that matter in any setting, is extremely uncommon. Thus, most box turtles killed by automobiles are breeding-age adults. This is likely to significantly affect the resilience and long-term persistence of local populations.

Fortunately, most pineland herpetofauna lack the aesthetic qualities that appeal to potential owners of a captive reptile. The Gulf Coast Box Turtle is an exception. Lacking great rarity and being a geographically restricted race of an otherwise common and widespread species, the Gulf Coast Box Turtle holds little appeal for serious hobbyists. However, the traits common to box turtles in general—hardiness, moderate size, an inquisitive and bold nature—make the Apalachicola subspecies a prime target as a novelty pet.

Although undemanding as a captive in comparison to many other turtle species, box turtles still require certain components in their environment that are frequently overlooked by the laypersons who typically own them. Box turtles are omnivorous and will consume virtually any palatable human food offered, leading some pets to become malnourished. A diet composed of a changing variety of fruits and vegetables, supplemented with a lesser proportion of protein in the form of earthworms, insects, whole flesh items such as chopped minnows, and small amounts of fortified canned cat or dog food, is necessary for long-term good health. Equally essential and inextricably intertwined with good nutrition is frequent regular exposure to ultraviolet radiation—nothing beats natural unfiltered sunlight. Sunlight is something many turtle pets are denied because they are kept indoors for all or most of the year, and artificial lamp substitutes are expensive and often bypassed. Lacking these basic components, along with clean quarters, fresh water, and proper temperature, a captive box turtle will appear healthy for a time, then begin to decline and eventually sicken and die, or sicken to the point where its owner releases it, often in a completely foreign and unsuitable location.

In recent years, conservationists have been alarmed by the massive consumption of wild turtle populations in Asia, both literally as a staple food, and figuratively, as novelty pets in more affluent quarters. While these activities were initially a threat only to local species, an inevitable exhaustion of these resources in concert with increasing demand has caused the focus to shift to native U.S. turtle species. Sliders (*Trachemys* spp.), Common Snapping Turtles (*Chelydra serpentina*), Alligator Snapping Turtles (*Macroclemys temminckii*), softshell turtles of various species (*Apalone* spp.), and box turtles have borne the brunt of this insatiable appetite. In one typical case, several consignments of Three-toed Box Turtles, *T. c. triunguis*, from northern Mississippi, approximately 150 individuals, were collected and sold for this purpose. The harvest of wild box turtles at that level is unsustainable at the local level, but when the taxon involved is as widely distributed and common as *T. c. triunguis*, such events may have little overall impact. For a more geographically restricted race like the Gulf Coast Box Turtle, the impact of a few seasons of commercial harvest could be devastating, thus the Gulf Coast Box Turtle, while not presently imperiled, is a vulnerable form.

The growth in international trade in box turtles during the early 1990s is shocking. In 1990, 3,000 declared exportations destined for Europe and Japan were recorded (Hoover 2000). This number probably reflects the annual tally for the preceding decade or more. In 1991, the number exploded to 13,585, then almost doubled to 26,361 the following year; in 1993, 23,420 box turtles, most if not all wild-caught adults, were sent to overseas markets (Hoover 2000). Thus in just four years, a staggering 63,366 box turtles were collected from the wild. These numbers represent only legally exported turtles. Although restrictions on such activity were virtually nil at the time, they still required that certain conditions be met and paperwork be filed, time-consuming processes that some individuals undoubtedly sidestepped, so the true numbers are higher and will never be known.

In November 1994, as a result of alarm in the conservation community over the numbers presented above, the United States petitioned the Convention on International Trade in Endangered Species (CITES) to list all races of the North American box turtle in Appendix II of the CITES protected list. This would require export permits be issued by the U.S. Fish and Wildlife Service for any exportation of this species, thereby allowing accurate monitoring of the trade, prosecution of illegal activity, and the ability to limit export numbers to annual quotas if deemed necessary.

Summary

Terrapene c. major is a unique form, yet little appreciated as a result of being nested within a diverse, polymorphic taxon. It is somewhat lost to common perception owing to the variety of *Terrapene* colors and patterns exhibited across the United States, but none of these are so closely associated with a particular habitat and locale. It is the most phenotypically distinct member of its species taxon, and our sketchy understanding of its natural history suggests a noteworthy ecology as well. Among the four eastern subspecies of *T. carolina*, *T. c. major* is the most habitat and region specific.

The two chief threats to wild populations, road casualties and commercial harvest, are not common to most other pine woods herpetofauna. The daunting problems threatening most flora and fauna of the southern pinelands—habitat disintegration and degradation—have yet to exert much impact on the Gulf Coast Box Turtle, judging from the appearance of its haunts. Thus the box turtle of Apalachicola is a special case for those concerned with the loss of pine woods biodiversity.

Brownchin Racer
Coluber constrictor helvigularis

Most southern pine woods herpetofauna live the precarious existence of an ecological specialist occupying an insecure environment. Their pineland habitats have shrunk and been shattered into a disconnected pattern of remnants on which degradations continue to occur. The great majority of taxa treated in this book have shrunk into relict distributions since their descriptions were originally written, and they currently inhabit a fraction of the land they once occupied. The ranges of only a minority of pine woods forms have escaped significant impact by commercial and residential development or land management practices that depart from Nature's regimes.

The Brownchin Racer is one of the few pineland endemics for which the expanse of suitable habitat has remained stable to the present day, and a combination of factors appears to be the reason. One of these is the nature of the snake itself. More so than its pine habitat conspecific, the Tan Racer, *C. c. etheridgei*, *C. c. helvigularis* retains the adaptive generalist traits that characterize most racers. Thus, although restricted to the mesic flatwoods of the Apalachicola River basin, it tolerates a broad range of conditions. Brownchin Racers are as likely to be encountered in a backyard woodpile in one of the small coastal communities of the region as they are in the most remote interior tracts of Tate's Hell State Forest. In keeping with the fundamental qualities of its genus, *C. c. helvigularis* is a frequenter of forest/clearing ecotones; thus the biotic edges created by industrial forests, residential lots, and roadways, as well as those naturally occurring, yield an abundance of shaded-sunny zone boundaries that provide good habitat for this snake.

All but the most adaptable of snakes require sizable areas of landscape left undeveloped enough to maintain functioning food webs and ecosystems, if only as a source of individuals for infiltrating and inhabiting more disturbed sites where survivorship is reduced. The general area where the Brownchin Racer occurs is blessed with an unusually large amount of acreage set aside by federal, state, or private protection from development. Apalachicola National Forest, Tate's Hell State Forest, St. George Island State Park, St. Joseph

Bay State Buffer Preserve, St. Vincent National Wildlife Refuge, and considerable expanses of privately owned but undeveloped rural tracts dominate the map of the area and create a well-linked assembly of wildlands where racers and other pineland wildlife have been assured survival. The largest concentration of human activity in the region, the City of Apalachicola, is a small community that functions gently against the backdrop of the pine woods that surround it. This is one of the few regions within the southern pine belt where the mortality risk to a snake is still dominated by natural dangers such as predation and disease.

Description

Adult coloration is distinctive both within and without its genus. The scales display a velvety sheen unlike any of the other medium-sized snakes in the area. The ventral color is also dark, ranging from black to very dark gray. This coloration is shared with several other eastern races of *Coluber constrictor*. The only defining physical character that distinguishes *C. c. helvigularis* from conspecifics is the coloration of the labials for which the snake is named. In adult specimens, the labial region is pigmented a deep, mahogany brown color, in contrast to other dark, unpatterned eastern racers in which these scutes are black or dark gray with white mottling (plate 4.7).

Juveniles are strikingly different from adults, being conspicuously blotched with dark markings on a light gray or tan ground color.

Distribution Notes

The Brownchin Racer has a very restricted distribution, encompassing the Apalachicola basin. The phenotype is seen in southeastern Calhoun,

southwestern Liberty, and western Franklin counties and the entirety of Gulf County, Florida. Outside this mesic ecozone, racers exhibit a rapid transition to a form assignable to the Southern Black Racer, *C. c. priapus*, but a narrow intergradation zone occurs in some sections of the interface between these two subspecies. *Coluber c. helvigularis* is abundant on St. Vincent Island in Apalachicola Bay.

Habitat

Low, wet, pine flatwoods are the typical situation where the Brownchin Racer is found. Interspersed throughout this association are smaller enclaves of drier habitat, but true ridge associations are uncommon within the distribution. This snake is probably associated with flatwoods more by necessity than by design, because it does not shun any sort of habitat that is conducive to snakes. It is more likely that the relative greater abundance of racers in Apalachicola flatwoods is the consequence of that habitat's ubiquity rather than any strong ecological preference on the part of the snake (plate 4.8).

Unlike other herpetofauna with which it is sympatric, the Brownchin Racer is as frequently seen in the populated areas lining the coastline as it is in the more remote forests to the north. At the present time, the communities from Indian Pass eastward to Alligator Point are undergoing a rapid transformation from rural, sparsely populated fishing towns to crowded beachside developments. This process has created a heterogeneous landscape of patchwork clearings, disheveled construction sites, and small tracts of natural pine woods. Such variation, with abundant ecotones and open spaces, is exactly the sort of situation that favors racers. As this process continues, the region will no doubt resemble the older coastal developments its designers aspire to, such as Panama City lying approximately 64 km to the west. That scenario is one fraught with uncertainty for the Brownchin Racer.

Ecology and Natural History

The Brownchin Racer is active throughout the year, seeking surface cover during winter cold snaps, when it can still be found by turning objects such as boards and tin. Like all *Coluber*, it is an active, visually oriented hunter of a wide variety of vertebrates and invertebrates. Its activity frequently leads it to roads, where it may pause to bask, but its alertness and speed prevent it from being struck by passing cars as frequently as many other species of diurnal snakes.

Our knowledge of the reproductive biology of this form is based on the observations of Herrington (1974) of a single specimen collected when gravid. Oviposition took place 4 days after capture on May 25, with 15 eggs being laid. The eggs had a coarse granular surface and were nonadherent, which are characters shared by all members of the genus. In fact, in all the notes provided by Herrington, no feature distinguishing *C. c. helvigularis* from other eastern subspecies was described.

Conservation

The Brownchin Racer is abundant within its limited distribution. In contrast with the other *Coluber constrictor* that is a pine woods endemic, *C. c. etheridgei*, *C. c. helvigularis* lives in territory that is composed largely of national or state forest, or private reserves, and is in no present danger of being forced into shrinking habitat. *Coluber c. helvigularis* occupies roughly the same expanse of territory that it always has. It is not a form that requires pristine wilderness, and it frequents areas along the margins of small towns scattered through the Apalachicola region.

Summary

Coluber c. helvigularis is the least specialized of the southern pine woods herpetofaunal endemics, being only weakly discernible from conspecifics found elsewhere in Florida and the Deep South. It appears not to have been shaped to any appreciable extent in habits, physiology, or morphology by the pine woods. This observation leads to an interesting question to conclude the account of this snake: what features or dynamics in the environmental or genotypic history of this snake lead to the fixed coloration trait that defines it? For now the question cannot be answered.

Apalachicola Lowland Kingsnake
Lampropeltis getula meansi

On a tiny island off the southeastern coast, a kingsnake prowled across the dunes in search of prey. Its home was one of a trio of islands, tiny spits of sand only a few kilometers long and barely high enough to maintain themselves as dry land against tides and stormy seas. The islands supported significant areas of mature forest at their highest and driest ends, but over most of their landscape the dominant flora were sea grasses, salt water–tolerant ground covers, and cacti. Vegetation was scarce, with most of the terrain being sun-baked dunes or sparsely treed scrub. The sunlight reflecting off the white sand was nearly blinding, and there was little cover where the 1.5-meter kingsnake could find refuge.

Although this snake had lived on this particular island for the past five years, it began its life on one of the neighboring islands, the smallest of the three. Less well vegetated than the other two islands, and so low in elevation that ephemeral pools created by drenching rains quickly became poisoned by the saltwater that seeped in from the high water table, the tiny islet had few vertebrate species in residence. Only a few hardy reptilian species, and birds that could come and go, lived there. Mammals were absent except as occasional strays that happened to reach the shore by swimming. The island had experienced a few periods where rats had managed to establish a breeding population, but these had been only transitory, and the harsh environment soon vanquished their brief reigns.

During its first three years, as it grew, the kingsnake fed on the abundant diurnal teiid lizards that thrived in the hot and xeric environment. Then, just as it was approaching adult size and beginning to have difficulty extracting sufficient nourishment for its growing length from an exclusive diet of scrawny lizards, a tropical storm passed over the island, sweeping away large chunks of debris with it. It was on a pile of this debris, a knotted tangle of palm trees, that the kingsnake was transported from the little island of lizards and deposited fortuitously on its present home, the largest and highest of the three, with several permanent freshwater ponds and an abundant rodent population. Here the kingsnake began to thrive again, and it was now a beautiful and impressively large example of its kind.

Such interchanges between the kingsnakes of the three islands had been commonplace for centuries, because of storms or simply because snakes would occasionally brave the shallow and normally calm straits that separated them and swim to a neighboring shore. Thus, all these snakes were interrelated and resembled each other in every respect. Not so, the mainland population of kingsnakes. They resided in a very different environment and were separated from the little islands by 16 km of wave-capped and current-driven sea. Any snake that attempted to make the trip, from mainland to island or vice versa, succumbed before succeeding.

In pattern and color, the island kingsnake was admirably adapted to survive in its sun-bleached environment. Unlike its close relatives on the mainland, which were richly colored in deep mahogany browns and jet blacks, and boldly patterned in dark blotches framed by lemon yellow borders, this product of the dunes was as pale and bleached as the driftwood littering the beach. Although each scale was half black, the other half was a pale straw gray, very similar to the off-white sands on which it lived. The cumulative effect was that of an overall pale yellowish gray snake. When it froze in mid-crawl upon detecting a possible threat, it almost vanished against the sand.

This scene, or one very much like it, played out along the coast of the Florida panhandle, but not in our lifetimes. Those who have made careful study of the peculiar kingsnakes of the low, wet pinelands of the Apalachicola region of Florida believe that the remarkable and uniquely patterned examples inhabiting the eastern portion of the Apalachicola lowlands are remnants of a day when three offshore islands shaped the kingsnakes trapped upon them into well-adapted phenotypes for their situation. A scenario like the one just described could have been observed 18,000 years ago. Today, with their island homes having coalesced almost imperceptibly into the mainland after sea levels fell, they appear as oddities, an enigmatic enclave of strange looking snakes with no obvious explanation for their existence.

Description

This race of kingsnake is marked with a variable but always distinctive pattern. Adult appearance can be divided into three basic pattern types: (1) very pale tan or grayish tan, with 2–25 even paler broad blotches, (2) very pale, with longitudinal striping, and (3) completely unpatterned, with every dorsal scale uniformly and sharply bicolored, imparting a spotted appearance. Most specimens are of the blotched phase, with a typical individual displaying roughly a dozen faint blotches separated by 70–80 lighter scales. Blotched snakes bear a resemblance to certain other races of Florida kingsnakes that occur in more general habitats, in particular those from the extreme southern part of the state. These "South Florida Kingsnakes" were once designated by the subspecies *L. g. brooksi*, but are now considered a phase of *L. g. floridana* (Blaney 1977). Striped specimens have the darker pattern elements oriented to produce a single longitudinal stripe running along the mid-dorsum, 1–3 scales in width. They are reminiscent of striped snakes of any number of species that occasionally appear from captive breeding projects when temperatures have exceeded normal parameters during embryogenesis, but in the case of these *L. getula*, the trait is apparently based on genetic, not environmental variations (Means and Krysko 2001). The most dramatic and divergent polymorphism is the patternless phase. As Means and Krysko (2001) point out, there are no other naturally occurring populations of patternless *L. getula* in the eastern United States. The extreme south Florida population ("*brooksi*") appear on gross examination to be uniformly pale straw yellow, but under good light can clearly be seen to be banded, albeit faintly (Means and Krysko 2001). The coloration of individual dorsal scales in the eastern Apalachicola lowland specimens is interesting and especially striking in patternless snakes. Each and every dorsal scale is sharply bicolored, the posterior half being darker than the anterior. With the light anterior half of each scale framed by darker pigment on all sides (by its own dark half posteriorly and by the dark posterior halves of two bordering scales anteriorly) each light area becomes a light spot, and the overall appearance of these snakes is a symmetrical speckling (plate 4.9).

Ventral pattern in these kingsnakes is also diagnostic. Each ventral scute is marked along its posterior edge with dark pigment, but otherwise unmarked in most specimens, resulting in a pattern of alternating light and dark narrow crossbands on the belly. Some specimens also have staggered pairs of squarish blotches running along each side of the ventral scutes, 3–4 ventral scutes in length and spaced 6–10 ventrals apart, probably reflecting an intergrading influence from surrounding populations.

Distribution Notes

Kingsnakes exhibiting all the characters of the relict barrier island race are encountered throughout Franklin County, the southern half of Liberty County, and unpredictably in central Gulf County, in the Apalachicola region of the Florida panhandle. The genetic influence of the taxon extends further out of this core area, most obviously toward the western division of Apalachicola, where intergrade patterns dominate.

Although an Apalachicola subspecies of *L. getula* has been included in the literature since the 1950s, the polymorphic Apalachicola Lowland Kingsnakes have only recently been recognized as a distinct evolutionary lineage. Neill and Allen (1949) described *L. g. goini* as a new subspecies restricted to the Apalachicola basin area west of the Apalachicola River. Widely spaced, long oval blotches characterize these snakes, and their distinctive pattern was unlike anything seen in the species at the time of their description. The validity of this taxon was questioned and determined to be invalid by Blaney

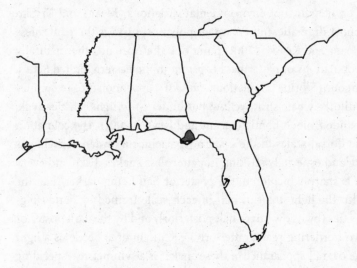

(1977), who considered it an intergrade between prehistorically adjacent populations of *L. g. getula* and *L. g. floridana*. Krysko (1995) first reported the presence of a recognizable and discretely ranging race of *L. getula* in the lowlands east of the Apalachicola River. He contended that although Blaney was correct in relegating *L. g. goini* to intergrade status, it was not the product of an ancient gene flow between *L. g. getula* and *L. g. floridana* but, rather, a more modern and ongoing character blend between *L. g. getula* and the newly discovered population in the eastern half of the bioregion. Additional work by Krysko (2001) followed with elaboration by Means and Krysko (2001) supporting the definition of a subclade relationship between Florida's *L. g. getula*, *L. g. floridana*, and the eastern Apalachicola Lowland Kingsnake. Krysko and Judd (2006) formally described this race as a subspecies of *L. getula* with the new name *L. g. meansi*, citing the fact that the name *L. g. goini* designated individual hybrids between *meansi* and *getula* populations. According to international rules of taxonomy, names applied to hybrids are designating multiple taxa, and as such, are not available to be used again should perceptions change, as in this instance.

Habitat

The dominant condition of the natural areas remaining is mesic pine flatwoods, originally longleaf or, along the coast, sand pine. Interspersed throughout are several major rivers and numerous secondary creeks, sloughs, and swamps. Kingsnakes favor the wetter sites within this landscape. Paradoxically, they are seldom found on the drier relict barrier islands that now present as sandy high ground, their ancestral home. The snakes utilize a variety of microhabitats, including bogs and salt marshes (Enge 1997; Krysko 2001). The two primary areas that envelop most of this habitat are the Apalachicola National Forest and Tate's Hell State Forest. Most specimens that have been found were on roads, often injured or dead from passing traffic, so very little is known about their habitat preference, and the best that can be assumed at present is that it differs little from that of conspecifics found elsewhere along the Gulf and Atlantic coastal plain (plate 4.10).

Ecology and Natural History

Surprisingly little is known about the natural history of kingsnakes. Virtually nothing has been published regarding the habits of the eastern Apalachicola Lowlands form, but in most respects they likely resemble details reported for other Florida populations. Kingsnakes living farther south on the Florida peninsula exhibit two seasonal activity peaks. The most common time to find south Florida kingsnakes moving about in the open is during March and April (Krysko 2002). This period coincides with breeding activity, and most specimens found are males. Another peak occurs in December, possibly correlated with optimal surface temperatures, and females are more frequently encountered then (Krysko 2002). In the Apalachicola flatwoods, the more northern clime may alter this pattern by delaying the spring peak into April and May and changing the late season peak to sometime in November.

Kingsnakes occurring in the southern coastal plains are chiefly diurnal as adults (Gibbons and Semlitsch 1987), although during very hot weather they also prowl at night and can be found crossing rural roadways. In contrast, juveniles are typically nocturnal, and most living specimens of juvenile Apalachicola Lowland Kingsnakes have been found crossing roads at night.

Conservation

Land development and ensuing habitat reverberations in the Apalachicola basin, a region that is rural and sparsely populated, have coincided with a measurable reduction in the abundance of kingsnakes. Krysko (2001) reported numbers collected during the 1960s and 70s throughout Florida, based in part on data tabulated by pine woods herpetologist D. Bruce Means. During the 1970s, 69 *L. getula* were documented from within counties where the eastern lowland race occurs. The total dropped to 10 in the 1980s, and Krysko knew of only one specimen collected during the 1990s. While not a direct reflection on the rate of decline in natural populations (because of the unscientific sampling; Krysko described the recent increase in collections following a change in sampling method), it strongly suggests a decrease in numbers in at least some areas of former abundance. Krysko recalled being able to observe several specimens crossing roads on days when conditions were right in the 1970s, and contrasted this with the extreme rarity of finding even one specimen in the same areas today.

The informally documented decline in kingsnake encounters is correlated with several factors that are well known to threaten the herpetofauna of the pine woods. Local residents and long-time visitors agree that there has been significant development in the Apalachicola wilderness in recent years. Commercial logging, on both private and public land, and modifications of drainage patterns have changed the character of some parts of the forest and fragmented other tracts. Long overlooked by developers while the rest of coastal Florida's leisure and residential infrastructure was established, Apalachicola is now getting their attention. Property values are rising, beachfront developments are springing up, and the attractiveness of this area for further exploitation is being enhanced.

Another potential factor in the decline of eastern Apalachicola Lowland Kingsnakes is the taxon's appeal in herpetoculture. *Lampropeltis getula*, with its myriad of attractive geographic variants, hardiness in captivity, and docility, is a staple in the pet trade and among reptile enthusiasts. Rare and distinctive phenotypes are the most sought after targets, commanding the greatest prestige for their owners and the highest prices. The boon in American herpetoculture that began in the early 1980s (de Vosjoli 1994) was preceded by an increase in demand for wild-caught specimens, as herpetoculturists built collections and breeding groups. The ensuing legion of those desiring to possess rare reptiles made it commercially rewarding for local collectors to look for rare species. The current understanding of geographic variation in Apalachicola kingsnakes and its probable origin was lacking during the 1970s. Consequently, the most desirable phenotype was the one then designated *L. g. goini*, or what is now considered an intergrade between the historical lineage inhabiting the eastern lowlands and populations outside the Apalachicola bioregion. Many specimens of the *goini* morph were harvested off the roads crisscrossing the western basin flatwoods, but how many is unknowable.

Eastern Apalachicola Lowland Kingsnakes face another threat that is peculiar to mesophilic flatwoods specialists. Because of the swampy character of the landscape, dredging and drainage pattern alterations are common practices employed by those wishing to utilize the land. Deepening of slow-moving creeks to limit the expanse of wetlands or the excavation of new channels to expedite drainage and eliminate soggy ground are changes that have wrought harm to most pine woods amphibian fauna; among endemic reptiles, no species is more sensitive to these changes than *Lampropeltis getula*, which throughout its large range in eastern North America is a frequenter of mesic habitats. Mesic flatwoods situated in less desirable locales than the Apalachicola coastline are not always subjected to these alterations, but now that interest in Apalachicola as a recreational and residential area is starting to gain momentum, the region is certain to evolve and be transformed from its natural state, and the local kingsnake, itself a product of a unique evolution, will become increasingly ill adapted and rare.

Summary

Although the survival of most pine woods herpetofauna is in jeopardy to at least some degree, not all of these situations are the result of mankind's activities. Some species are whispers of a prehistoric time, when the earth was a very different place. As the landscape evolved, many species of reptiles and amphibians changed, eventually disappearing as their progeny evolved in response to the altered environment. A few species still remain with us, true "living fossils." The distinctive color and pattern morphs of kingsnakes in the Apalachicola region of Florida are an example of this kind of natural extinction process. Although human disturbances in their habitat have hastened the pace, these snakes are probably heading inexorably toward being subsumed by neighboring phenotypes that now surround them on all terrestrial fronts.

5

Savanna Specialists

Dusky Gopher Frog
Lithobates sevosus

The word "endangered" has been carelessly used when describing biological circumstances. So variable are the situations that get labeled endangered, the term has lost much of its ability to convey a sense of conservation urgency. Most famously employed as a designation by the U.S. Department of Interior to categorize the plant and animal species in most imminent danger of extinction, the term imparts an objectively derived and standardized level of population imperilment when used in this way. Such was the situation with the Dusky Gopher Frog, *Lithobates sevosus*, in 2000, when the species was officially so labeled. An Endangered taxon is one for which a set of circumstances have rendered it vulnerable and inadequately defended against identified threats. To be so designated, these forces need not be currently influencing the biological population in question if the taxon is without apparent defenses or resiliency to prevent the dire consequences of such threats. By the turn of the millennium, the Dusky Gopher Frog had reached such a point. It was known from only one site, a small portion of De Soto National Forest in Harrison County, Mississippi, centered around a small ephemeral pond known as Glen's Pond, named in recognition of the U.S. Forest Service biologist who discovered it. Viewed at a local scale, and ignoring likely future events, the Dusky Gopher Frog population was small, isolated, but secure. The habitat within the few hundred acres of National Forest surrounding Glen's Pond was nearly optimal, being mature second growth longleaf forest, expertly attended by U.S. Forest Service managers. The pond itself was a rare pine woods jewel, a landscape feature virtually extinct, but preserved in its finest form in this small corner of timberland. The catchment was a gently sloping basin, more of a low spot on the ground than an actual pond. Its substrate was sand, not clay, and as a result of frequent burns the emergent vegetation was grasses, with very few woody shrubs or sapling hardwoods, when the pond filled each winter. Because of the inorganic substrate and lack of leafy litter, its water, during the few months of the year when it held any, was clean and clear, acidified and tinted slightly by the pine straw that lay on the bottom. This was a perfect site for the 50-odd *L. sevosus* that inhabited the surrounding pine woods, the last of their kind, to visit each fall and reproduce into perpetuity. In fact, some contended that the gopher frogs had their home there at Glen's Pond, and needed no further protection. If nothing were to change, this might have been true.

Many populations of plants and animals are continually challenged by environmental changes. Some of these perturbations come in the form of natural events such as floods that physically sweep away individuals, storm damage that fells trees and alters forest structure, and the dynamic changes, not always favorable, wrought by fire. Other threats are the ones more commonly lamented by conservationists, the human-caused changes such as the clearing of forests, conversion to agriculture or grazing land, development for commercial or residential use, and the fragmentation of large tracts by an expanding roadway network. Species comprising large, interconnected populations spread across a large area are usually able to weather these changes as a whole, even if local extinctions occur. However, when a species has degenerated to the point reached by the Dusky Gopher Frog by the spring of 2003, any of the threats listed and a host of others, some unforeseeable, can quickly exterminate it entirely. This was the scenario that unfolded for *Lithobates sevosus*.

In the early spring of 2003, the reproductive season for Dusky Gopher Frogs was well under way at Glen's Pond. A cohort of thousands of nearly full-grown tadpoles from a late autumn spawn inhabited the pond, and a second spawning event had recently contributed thousands of additional prospects for recruitment into the dwindling population. All things considered, it looked as if it was going to be a good year for the Dusky Gopher Frog. Then, abruptly and without prelude, all the larvae began dying. The tadpoles exhibited symptoms of illness—bloating, lethargy, floating at the surface—and the scientists soon discovered that the sick larvae were infected by a protozoan-like parasite that they had not seen before. The rapidity with which the pond transformed from one teeming with frog life to a sour pool stinking with the stench of mass die-off was shocking to the biologists who had been diligently monitoring the frogs, who now witnessed progress of the decimation day by day. In the end, only a few hundred larvae survived, most of them part of a group that had been collected as egg masses and raised out of the pond in cattle watering troughs.

The story's end has yet to be written. Conservation biologists are trying to understand the disease's etiology and thwart further epidemics, but whether anything can be done at this point is uncertain. What is certain is that the situation faced by the Dusky Gopher Frog is not unique in the sense that several other pine woods reptiles and amphibians are equally susceptible to the unpredictable forces that can unleash a conservation crisis.

Description

With its squat, toadlike physique and rugose skin, *Lithobates sevosus* and the closely related Crawfish and Carolina Gopher Frogs (*L. areolata* and *L. capito*) are the most morphologically divergent of North American ranids, a family in which the basic bauplan is only subtly varied among species. *L. sevosus* has relatively short legs and tends toward a dumpy physique, less streamlined and athletic in appearance than other members of the genus. The skin on the dorsum, flanks, and limbs has a pebbly texture, owing to an abundance of wartlike glandular bodies. In fact, the overall appearance is more reminiscent of a toad than a typical frog.

Some specimens are quite dark in color, and the lack of contrast between the ground color and pattern results in an overall "dusky" or "dark gopher frog." Other specimens have pale gray or greenish gray ground color and are boldly marked. This variation is genetic, as individual gopher frogs show little capacity to change color with fluctuations of temperature or light. Venters are mainly unpatterned and off-white in color, but individuals may have varying amounts of faint gray clouding or blotching on the gular and sternal regions (plate 5.1).

Larvae are pale tan with metallic gold highlights on the dorsum, especially atop the head. As they grow, traces of pattern, at first barely perceptible, become clearer so that toward the end of larval development they are fairly easy to identify by the distinctive adult pattern. However, reliable diagnostic pattern or coloration characters that can distinguish *L. sevosus* larvae from sympatric Southern Leopard Frogs, *R. sphenocephala* are not remarkable during most of the larval period (plate 5.2).

Distribution Notes

The present-day distribution is restricted to two sites in Harrison County, Mississippi. Two apparently disjunct populations are represented, one centered around Glen's Pond in De Soto National Forest, and the second near another ephemeral rainfall catchment dubbed Mike's Pond, where approximately 60 larvae were discovered by biologist Mike Sisson in 2004. Approximately 32 km of uninhabited territory separate the two sites. Historically, the species had a much larger range, encompassing the area between Mobile Bay, across coastal Mississippi, and into Louisiana's Florida Parishes (east of the Mississippi River and north of Lake Pontchartrain). Although no confirmed records have been made outside of Harrison County in decades, a search for other relict populations is underway and it is unlikely but possible that additional populations may be found in some small corner of the pineland fragments that remain in Alabama or Louisiana.

Habitat

The general habitat as well as the specific features of the microenvironment that adults occupy for most of the year are endemic components of longleaf pine upland along ridges. This association, within the range of *L. sevosus*, is also favored by *Pituophis m. lodingi*, *Rhadinaea flavilata*, and, historically, *Heterodon simus*. However, these reptiles have broader tolerances of hydric variation than the amphibian fauna of the southern pine woods. A limiting factor for *L. sevosus* habitat is the quality and availability of aquatic environments during the winter breeding season. Since adult frogs don't move far from the breeding ponds during their lifetime (Richter et al. 2001), it is the temporal and physical characteristics of the aquatic portions of their territory that define the specific habitat of the Dusky Gopher Frog. These uncompromisable needs also explain the frog's absence from many tracts of otherwise suitable land. With these requirements in mind, it is reasonable to conjecture that the Dusky Gopher Frog has always been relegated to small, localized populations where fortuitous conditions permit the consistent formation of wintertime ponds in a very xeric forest matrix.

The Dusky Gopher Frog seeks breeding pools of a precise nature. The ponds must fill and remain full on a schedule that keeps them available during the appropriate season and provides a dependable refuge for larvae for the duration of the protracted developmental period. In porous-soiled sandy uplands, locations that can collect and retain water are much less common than in flatwoods or savanna associations. As for most anurans, suitable catchments must be filled by rainfall rather than with floodwater overflow from permanent bodies, and dry completely each year, to prevent the development of a fish fauna and to hamper the establishment of predatory invertebrates, further limiting the number of inhabitable sites (plate 5.3).

Glen's Pond offers a tutorial on the breeding requirements of *L. sevosus*. Frequent burns in this portion of De Soto National Forest have preserved the open midstory quality of good longleaf forest. The upland terrace supporting the pond and surrounding environs are grown to well-spaced pines that impart a thin, open canopy. This is the condition encountered by the gopher frogs when they arrive at the pond to breed. The emergent vegetation is abundant and consists primarily of the same wiregrass that carpets the tract when the dry basin transitions and blends into the surrounding forest landscape during midsummer. The sparse growth of pines and paucity of hardwoods leave the pond exposed to the sun continuously through the day (plate 5.4).

Calling it a pond creates the inaccurate perception that the catchment is a banked depression of significant depth. In reality the basin is so slight that it would be easily overlooked when dry. Maximum depth in Glen's Pond is a mere 1.1 meters (Richter et al. 2003) and much of its area never exceeds 30 cm (Mike Sisson, pers. comm.). Glen's Pond is actually little more than an expansive puddle, roughly 450 m around its margin when full, and gives no visual indication that it is one of the most critical conservation sites in the southeastern United States (plate 5.5).

Because the habitat utilized by *L. sevosus* is a combined consideration of two very different summer and winter characteristics, an otherwise optimal breeding pond is not acceptable to this species unless it is embedded in the correct broader environmental context. Conversely, good forest habitat lacking the preferred aquatic component fails to satisfy the needs of this amphibian. Particularly in modern times, this precise pairing of aquatic and terrestrial parameters is exceedingly rare. At Glen's Pond, a portion of the surrounding landscape is longleaf pine ridge. This area affords good summer refugia for gopher frogs in the form of burned-out stump holes and Gopher Tortoise burrows (Franz 1986), although the latter have declined sharply in recent years.

The hydroperiod in ponds within tracts of good gopher frog habitat must coincide with the winter breeding period of this species. Ephemeral rainfall catchments exist, within the historic range of *L. sevosus*, on pine ridges that either fail to dry consistently each summer and thus harbor an abundant fauna of predatory insects, or, lacking the correct substrate compaction and composition, cannot retain water of an appropriate depth for the duration of the developmental period of the frog. Precise qualities of vegetation, depth, seasonality and persistence, surrounding forest structure, and sympatric flora and fauna, all coalescing within a small geographic area, are the essential ingredients for predicting the presence of Dusky Gopher Frogs. When one considers the narrow requirements of the species and the unlikelihood that even a subset of these conditions will be found in union at one location, the delicacy of the Dusky Gopher Frog's status in the wild can be appreciated.

Ecology and Natural History

The breeding season of *Lithobates sevosus* is variable but occurs sometime from October through April during years when rainfall patterns support reproductive activity. Heavy rains sufficient for filling upland catchments trigger breeding migrations. Frogs in transit favor rainy nights, and periods of dry weather cause them to take refuge and cease travels until moist conditions return (Richter et al. 2001). The xeric environment that gopher frogs traverse is a challenge faced by few other southeastern anurans.

At least at their last remaining refuge, there is great variability in the time of onset and length of the hydroperiod (Richter et al. 2003). Some years, the spawning site never fills, or does not persist long enough to allow the larvae to develop, hence there are years without recruitment. Richter et al. (2003) found that during a 13-year period Glen's Pond filled as early as September but was typically not suitable for gopher frog breeding events until December, and during some years it did not fill until April.

Egg masses are distinctive from sympatric ranid frogs. The clutches are encased in a gelatinous binding that is firmer and more spherical than that produced by the Southern Leopard Frog, which utilizes the same breeding sites and at roughly the same time. Dusky Gopher Frog egg masses tend to be deposited in deeper sections of the pond than those of leopard frogs, at a depth of 25–35 cm (Mike Sisson, pers. comm.). The deposition of eggs in suitable and favored locations is a life history trait that requires the seemingly incongruous juxtaposition of persistent aquatic environments embedded within dry upland habitat. The water basin must not only be of sufficient depth, but also support a specific flora preferred by the frogs for attaching eggs. Only ponds that burn regularly as part of their dry season cycle, and thus are carpeted with wiregrass and small herbaceous plants, will serve as sites for egg deposition. The gopher frogs attach their eggs preferentially to such grasses. If the pond basin does not experience frequent fires, the entire character of the aquatic substrate is changed and *L. sevosus* will reject it.

It is difficult to ascertain the longevity of anurans in the wild, *L. sevosus* being no exception. Following individuals across the course of their lifetime via radiotelemetry is possible but costly and extremely labor intensive. Frogs thus burdened with a transmitter, even one of the smallest models currently available, experience higher rates of mortality than unmarked individuals. Richter and Seigel (2002) estimated annual survivorship of *L. sevosus* by marking frogs as they scrambled out to the surrounding woods after metamorphosis or breeding, and documented the return of individuals over ensuing years. Based on the return rate for individuals, a survival rate of two consecutive years was estimated at only 15–17 percent, a startlingly low figure. Under normal conditions, where multiple nearby ponds would be present and individual frogs could emigrate to other sites, such an estimate might err on the excessive side, but since no other suitable habitat exists within reach of the gopher frogs living at Glen's Pond, Richter and Seigel's dire picture of a rapid turnover rate of individuals is probably accurate.

Adding to the difficulties faced by this lone population is the low rate of reproduction per individual. An unusual characteristic of *L. sevosus* reproductive biology, at least at Glen's Pond, is that these frogs behave much like annual plants in terms of reproduction. The typical Dusky Gopher Frog emerges as a young metamorph in the spring. It then spends the next nine or ten months, should it survive this tender period, in its upland habitat, sheltering from the xeric conditions, feeding during the milder conditions of night or during overcast or rainy days, and growing. When the pond fills again the following winter, assuming rainfall is abundant that year, the frog, now barely one year old, makes its way across the sandhills to spawn, never to return to the pond again. What becomes of these individuals after reproducing is not known with certainty, but with survival odds of barely better than 1:10, most are presumed to perish before the arrival of the next opportunity to breed.

Conservation

Amphibians of many species persist as a whole while local populations arise, thrive for a time, then collapse and vanish. If one could observe the entire geographic distribution of a species from some aerial vantage point and perceive each local population as a pinpoint of light, a uniform field of illumination would be seen, but if filmed and played back at time lapse speed, the distribution would present as a twinkling field of stars, with individual points extinguishing as new lights appeared. This dynamic process would be more or less in equilibrium so that the total expanse of territory would remain occupied by the species. The gopher frogs appear to have historically been such a species, with high rates of local extinction but in balance with the genesis of new populations as individuals emigrate to other good sites. Because of the harsh conditions that distinguish their preferred habitat and the inconsistency of optimum microenvironmental conditions, especially with regard to spawning sites, the long-term persistence of any local population is uncertain and, for many, probably brief. This is the balance of nature in play, and was a dynamic impinging on the species long before the activities of mankind began to have their own powerful impact on the habitat. So long as there are abundant sites that harbor small populations and avenues for gene flow between them, the loss of any single population has little or no impact on the stability of the species as a whole. Unfortunately, the precarious status of *Lithobates sevosus* precludes any resilience if the last few occupied sites are disrupted. Although at least a few breeding individuals inhabit Mike's Pond, and the possibility of other, undetected pockets of occurrence cannot be ruled out completely, the loss of the 50 or so individuals remaining in the environs of Glen's Pond would be the last straw for this species.

The waxings and wanings of the water levels in Glen's Pond through the turn of the seasons each year could hardly be less supportive of its most loyal resident species. Since the majority of frogs return to breed only once during their lifetime, a year without juvenile recruitment because of inadequate water levels or persistence puts the onus of population survival on the small minority of frogs that are able to return in a successive year for a second spawning attempt. With a repetition of an unsuccessful breeding year for a second consecutive season, only a handful of hardy individuals will still be alive and potentially capable of producing replacements for the aging and dwindling population. With three consecutive breeding seasons missed or disrupted before metamorphosis, the frogs at Glen's Pond will probably disappear within a year or so, as the reproductively senescent survivors die off. During the time the pond was under intense scrutiny in the years leading up to its Endangered Species listing, a two-year span occurred when no metamorphs left the pond (Richter et al. 2003). Richter and his colleagues looked at the probability of extinction of *L. sevosus* based on the fact that Glen's Pond harbors the only known population with any long-term viability, the natural lifespan of adults is three to four years, and most adults return to the natal site to breed only once or rarely, twice. They reasoned that if four consecutive years with no reproduction occurred, the population that remained would be beyond recovery and soon extinguish. During their long-term study, nearly three out of four years yielded zero juvenile recruitment, albeit not consecutively. They thus calculated the grave statistic of a 0.125–0.316 probability of the extinction of the Glen's Pond population of *L. sevosus*.

A new development has recently emerged that may be the final insult to the Dusky Gopher Frog. A housing and commercial development of 1,861.5 ha has been levied on land lying just to the north of the environs around Glen's Pond. The area is slated for a residential and commercial community hoped to house a population of 30,000 people, and construction was underway the summer that the first diseased larvae began their mass die-off. This activity has done two things likely to impinge on the gopher frogs. First, it has created a physical barrier in the form of open cultivated grass and asphalt over which the frogs cannot safely move, imprisoning them at the Glen's Pond woods. Second, although De Soto National Forest receives regular burning treatments and there is no direct reason this cannot continue, the extremely close proximity of the critical portion of the forest to new residential structures will doubtless hamper the future of this activity.

Summary

The southern pine woods are a harsh environment for any amphibian. Species adapted to these habitats are constrained by the paucity of moist refugia and dependable spawning sites. Even when the landscape is pristine and environmental conditions optimal, the Dusky Gopher Frog exists in a dangerous and stressful situation. The microenvironments sought by the frogs during the summer months—a tunnel created by a rotting root system at the base of a pine stump, or a Gopher Tortoise burrow—are safe havens. The gopher frogs are able to reach these sites only by traversing the sandy barrens during periods of rainfall. Once the frog is secreted within its moist shelter, it becomes totally dependent on the stability and soundness of the chosen sanctuary. Should weather, a marauding animal, or a human being disturb the summer refuge, the frog thus displaced is evicted into a deadly environment with little hope of finding a new location where it can hide before desiccation occurs.

The Dusky Gopher Frog, which has survived in spotty colonies under these trying conditions until recent times, now faces an array of new threats that appear to be overwhelming the species' ability to adapt. Its habitat, always disjunct and uncommon, has been fractured and quickly whittled away by man's activities. Its last refuges are jealously eyed and under intense attack by the continuing expansion of development. Just as the species slips away everywhere except for one small patch of pine ridge habitat, a devastating disease has struck. Despite the best efforts of conservation biologists, a dire but reasonable prediction can be made, that the Dusky Gopher Frog will be the first pine woods reptile or amphibian to become extinct in the wild, and that this day is not far off.

Mimic Glass Lizard
Ophisaurus mimicus

Momentous days in American herpetology come to the principals without warning. The first requirement for an historical event to occur on a given day is the presentation of an opportunity, which could come about as the result either of tenacious and dogged labor or, just as likely, of sheer good fortune. Perhaps these opportunities come more often than we realize, but the second essential element leading to an important herpetological discovery or insight is for an individual to be attuned to perceive the opportunity. How many great discoveries have fallen into the hands of scientists, only to be unrecognized and discarded, we can never know. A hint of the capriciousness of a fully consummated biological revelation is illustrated by the story of the discovery of the Mimic Glass Lizard.

During the 1950s, herpetologists working along the Gulf Coastal Plain began noticing that some of the glass lizards they were finding were slightly odd in appearance. At the time, three species of *Ophisaurus* were known, the Eastern Glass Lizard (*O. ventralis*), Slender Glass Lizard (*O. attenuatus*), and Island Glass Lizard (*O. compressus*). The puzzling specimens were encountered in disparate locations in Florida, Alabama, and North Carolina. They displayed subtle departures from scalation and pattern seen on the vast majority of Slender Glass Lizards from these areas, and were regarded as aberrant examples of that species. Although significant enough to draw published comment, they were otherwise preserved, cataloged, and deposited in museums in routine fashion, and not considered further.

One of these scientists was William Palmer, then curator of herpetology at the North Carolina Museum of Natural History. Palmer's first encounter with the strange glass lizards was in 1958, shortly after the time that E. H. McConkey had pondered some confusing specimens from Alabama and Florida. Some of McConkey's lizards had been collected and previously examined by the seminal Florida herpetologist team of W. T. Neill and E. R. Allen, who passed them along as *O. attenuatus*. By the mid-1950s, the U.S. herpetofauna was well known; although new species were occasionally named, this was usually the result of a reshuffling of boundary lines separating taxa as perpetually fluid species concepts shifted or new techniques of analysis or statistical tools discerned hidden distinctions within known assemblages. By this point in American herpetology, few locales had not been thoroughly collected by biologists at some time, and outright discoveries of novel creatures, especially among the more conspicuous vertebrate groups, were rare events not purposefully sought.

Palmer was certainly not prospecting for undescribed herpetofauna the day his path crossed that of the strange lizard's. The setting was a blacktop highway transecting low savanna habitat in rural Brunswick County, North Carolina. Palmer's attention was caught by an elongated object on the side of the road; as is standard for a biologist, he stopped his car and got out to investigate, roadkills of any kind being valuable sources of biogeographic information. He found a spoiling carcass of a glass lizard that he identified as *O. attenuatus*, but it was troubling on several key diagnostic points. Having no container or preservative with which to maintain the putrefied remains, he left it where it lay. Thus, what might have been the holotype of the fourth species of American glass lizard rotted into anonymity on a roadside, with no certainty that biologists would ever perceive this pine woods entity.

That specimen didn't enter Palmer's mind again until 12 years later. Bill was returning from a collecting trip where he and other staff of the North Carolina Wildlife Resources Commission had seined a river for fish. Accompanying him in the vehicle on the trip back was Robert Clark, a student and intern at the Wildlife Commission. Amid the idle discussion that waxed and waned during the long drive, there was an abrupt interruption as they spotted a wriggling form on the road as they passed through Bladen County. The characteristically stiff and frantic undulations of the creature were more suggestive of a legless lizard than a snake, and Bill, always on the lookout for new county records, thought this might be a significant distribution point for the Slender Glass Lizard. Bill noted that this lizard differed slightly in pattern from the Slender Glass Lizards he was familiar with, much as the roadkill specimen had a decade earlier. The lizard was caught and taken back to the museum for examination.

The exact moment of realization that a new species had been found is hard to pinpoint. More accurately, it began as a puzzlement that grew into a hunch, which led in turn to a quest for scientific confirmation. The epiphany occurred as Palmer began to examine specimens of *O. attenuatus* borrowed from other museums. He observed that the same suite of morphological and mensural oddities cropped up repeatedly in lizards from a wide range of localities. He soon realized that the "aberrant" glass lizards he and others had been finding along the Atlantic and Gulf coasts presented a discrete set of diagnostic characters. It was at this point that Palmer knew he had found a previously unrecognized species of glass lizard.

Description

The delay in discovering *O. mimicus* and the circumstances surrounding the event point to the similarity of this species with other glass lizards. Although there is no evidence that this species actually mimics congenerics, the resemblance of *O. mimicus* to *O. compressus* and especially to *O. attenuatus* requires close examination when identifying glass lizards throughout much of the Gulf and Atlantic coastal plain.

Several gross morphological features are shared by most species of glass lizards. The superficial similarity that these legless lizards share with snakes is immediately betrayed when one is handled. Glass lizards feel hard and stiff in comparison with the fluid, supple feel of snakes. Glass lizards usually thrash wildly when first seized, often causing the tail to break off. Mimic Glass Lizards, as is typical of most species, have fracture planes in the caudal vertebrae that make the tail quite brittle. Glass lizards have lateral skin folds running the length of the body (but terminating at the tail). These creases hide softer, finely scaled, flexible epidermis that allows the rigid trunk to expand when inspiring, after a large meal is consumed, and most importantly, when females become heavy with eggs. When identifying glass lizards, this fold is an important reference point for comparing elements of pattern that diagnose the species.

Three of the four species of glass lizard occurring in the southeastern United States, *O. attenuatus*, *O. compressus*, and *O. mimicus*, exhibit a striped pattern along the anterior flanks. Each species can be distinguished by the number of stripes and their position in relation to the lateral fold. Mimic Glass Lizards have 3–4 dark stripes above the lateral fold behind the head, and only vague or no striping below the fold. *Ophisaurus attenuatus* also has 3–4 stripes superior to the lateral fold, hence the close resemblance to *O. mimicus*, but it also displays finer, broken striping below the lateral fold, which is the best field mark to differentiate the two species. *Ophisaurus compressus* has but a single stripe above the lateral fold and can also be diagnosed by its lack of caudal fracture planes and thus a tail that is not brittle (plate 5.6).

Scalation can also be used to confirm the identity of a suspected Mimic Glass Lizard. In the region of the head between the mouth and the eye, a row of small scales (supralabials) directly above the scales that border the mouth (labials) will be noted. If the specimen in hand is *O. mimicus*, additional scales (lorilabials) above the supralabials prevent the latter from contacting the lower edge of the eye, while in *O. attentuatus*, the supralabials normally meet the edge of the orbit directly (Palmer 1987).

Ophisaurus mimicus is the smallest species of glass lizard in the United

States. In Palmer's series, one mature male had a snout-vent length of only 125 mm, and no specimen he examined exceeded 181 mm in body length.

Distribution Notes

The Mimic Glass Lizard occurs in suitable habitat throughout the Atlantic and Gulf coastal plains, from approximately Cape Fear, North Carolina, to Homochitto National Forest in southwestern Mississippi. Many reptiles and amphibians that occur in the pine woods of southwestern Mississippi are also found in the Florida Parishes of Louisiana (Washington, St. Tammany, Tangipahoa, Livingston, St. Helena, East Baton Rouge, and West and East Feliciana). Therefore it is probable that *O. mimicus* has inhabited Louisiana recently if not currently.

The species is not confirmed from peninsular Florida, but a single specimen raises some doubt as to its distribution in the state. Palmer (1987) noted one specimen labeled as having originated from Long Pine Key, at the state's southern extreme. This record should be considered with skepticism, as Palmer advised, although not discarded since the taxon is scarce and easily misidentified.

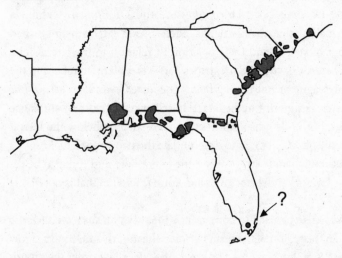

Habitat

The Mimic Glass Lizard is endemic to classic pine savanna. This association presents a wide range of moisture conditions through the course of a year. After periods of heavy rain, the ground becomes saturated and may remain so for long periods. Standing water may accumulate in the slightest of low-lying areas because of the level terrain and generally poor drainage. During drier weather patterns, savannas can become quite parched. Porous soils and a surface fully exposed to the sun prevent retention of moisture from intermittent light rain showers.

Based on the immediate environment at collection localities, Mimic Glass Lizards prefer or require exposed, sunny habitat. This condition is maintained by fire at inland locales, and by a combination of fire and storm damage along coastal stretches exposed to the full brunt of hurricane winds and tidal flooding. The Mimic Glass Lizard is the only native *Ophisaurus* so closely associated with fire ecology. Recent collections in Alabama have been made in the vicinity of bogs.

Ecology and Natural History

All native glass lizards are diurnal when active, but usually secretive. Commonly, they are discovered when objects such as old boards and roofing tin are lifted. Other specimens have been discovered as chance events, when an idle turn in a forest trail or country road has intersected the path of a glass lizard on the move. Experienced glass lizard hunters learn to recognize certain characteristics that identify roadways where these animals are likely to be found. Foremost, of course, is that they traverse good savanna habitat, but the most productive roads are those with shoulders open and grown to wiregrass, rather than those with forest abutting them directly. Perhaps the grass-bordered roads provide cover for foraging glass lizards along with the warm sunny conditions these diurnal lizards seem to prefer. Sunny mornings are the best time of day to observe basking glass lizards when road cruising.

Nothing is known of the reproductive biology of *O. mimicus* through direct observation. This is a poorly understood creature, and because of its secretive nature and ease of misidentification, details about its life history have remained as elusive as the animal itself. Gerald (2005) reviewed the literature pertaining to the reproductive biology in U.S. glass lizards and found no studies addressing *O. mimicus*. However, as a group, glass lizards are both uniform in natural history among species studied, and yet distinctive from other native lizard taxa in many respects, so generalizations based on observation of congenerics, especially the sympatric and morphologically similar *O. attenuatus*, probably paint a reasonably accurate picture of the Mimic Glass Lizard's reproductive characteristics.

Slender Glass Lizards from the Gulf Coastal Plain lay 5–10 eggs (Mount 1975) in late spring or early summer. The slightly oval eggs are nonadherent and measure approximately 1.5 x 2.0 mm. Observations of both *O. attenuatus* and *O. ventralis* consistently bear out that females coil around their clutch and remain with them until hatching. The purpose of this behavior remains a mystery, as it is unlikely these lizards would be effective in defending the eggs from most predators. Measurements of temperatures within and around the eggs provide evidence of very slight temperature differences possibly attributable to the brooding female (Gerald 2005). Until direct study provides a contradiction, we can assume that *O. mimicus* also exhibits this brooding behavior. Because of the small size of the species, the clutch size likely ranges lower in *O. mimicus* than for other species, probably not exceeding 8 eggs for the largest females.

Conservation

For species such as the Mimic Glass Lizard, so secretive in nature and poorly understood by biologists, it is impossible to know whether populations are stable or in decline. The Mimic Glass Lizard is so infrequently seen that decades could pass before its absence in an entire state would cause concern. Two conflicting hypotheses can be reasonably argued. One holds that *O. mimicus* continues to thrive in coastal plain pine savannas, rarely seen and usually not recognized, as has always been the case. The other is an alarming view that contends that the inability to repeat collections at some localities over the passage of several decades indicates a species in precipitous decline and already extirpated in many areas. At present there is no means to decisively judge these statements.

I believe that the Mimic Glass Lizard, while undeniably elusive even within strongholds that surely persist, is absent or artificially rare in locations where it was once abundant. This is an opinion I base on the fragmented condition of pine savannas conspicuously apparent along the Atlantic and Gulf coastal plains, and the alterations in the natural forest structure in many of these habitat remnants. The distribution map presented here represents the most optimistic assessment of the species' status, as it assumes the lizard occurs in all savanna habitats within the general area of documented specimens. Even this prediction indicates a disjunct distribution with small, isolated enclaves bearing particularly high risks of extirpation. The lack of knowledge of this species prevents us from realizing the intricacies in its ecology that may make it sensitive to certain disturbances that have already changed the character of many savannas. In virtually every instance where our depth of ecological insight into a particular pine woods endemic has advanced significantly, such species-specific vulnerabilities have come to light. We know that even when habitat conditions are suitable for supporting adult Pine Barrens Treefrogs, for instance, alterations in landscape drainage may impair seepage bogs and impact reproductive recruitment. Without the detailed knowledge we have of the reproductive ecology of the treefrog, such a statement could not be made and any efforts in preserving Pine Barrens Treefrog habitat would be insufficient. Biologists are still very much in the dark about what environmental components are critical for survival of the Mimic Glass Lizard, but indications that it favors bog habitats suggest that it shares the vulnerabilities of the Pine Barrens Treefrog. The Mimic Glass Lizard functions within a complex and nuanced ecological relationship with elements of its physical environment and a myriad of cohabitants. These still mysterious details are the key to understanding the species and conserving it.

Summary

Periodically catching the passing attention of herpetologists, *O. mimicus* remained unnoticed and wasn't tenaciously pursued until a specimen fell into the hands of a biologist whose thoughts and curiosities were properly tuned to receiving the new insight. Many of the reptiles and amphibians of the southern pine woods were likely caught and examined by biologists long preceding the individual who actually described the taxon. That is why all scientists should approach the routine tasks of their specialties with an open mind and optimistic attitude that much knowledge and understanding is still hidden. Biologists working in the southern pine woods must remain fully awake to the possibility that, on any given day, the most trivial of moments may lead to important discoveries.

Pine Woods Snake
Rhadinaea flavilata

We perceive and value organisms through the filter of our criteria for significance and utility, which often leads us away from appreciating so many beautiful creatures. This tendency characterizes biologists as well as the lay public. Three qualities highly regarded in Western culture and reflected in the wildlife we most value are large size, bright coloration, and the capacity to inflict harm upon us. With vertebrates, for instance, it is the small and secretive species, the modest inhabitants of the forest litter that are most poorly known and have the fewest dedicated workers on their behalf. Among the insects, everyone is familiar with the imposing praying mantids, katydids, and luna moths, or the gaudy monarch butterfly, yet who but an expert is even aware of the existence of the ubiquitous phorid flies, which are everywhere to be found if someone only seeks them? And who living in the South has not heard of the dangerous fire ant, and cannot identify its mound with merely a glance from a moving car, but to whom the legions of other common ant species and their lifestyles are completely unknown?

Similarly, the most familiar pine woods reptiles and amphibians are those that are large, gaudy, or venomous. Most people who are not otherwise interested in herpetology can nevertheless identify coachwhips, indigo snakes, and diamondbacks if they occur locally. These species are easily recognizable by virtue of their large size or ability to cause harm. Similarly, residents of the pine belt are aware of coral snakes and the various harmless snakes that mimic them, at least to the point where they can distinguish them as kingsnakes or milksnakes, because of the importance of being familiar with a potential danger and because of their garish coloration that draws the human eye.

The problem with this approach toward knowing the natural world is that it ignores such a multitude of organisms. The widely familiar herpetofaunal elements of the southern pine woods are only a small subset of the unique diversity that occurs there. Many of the endemic species are neither large, colorful, nor dangerous. Quite a few are humble, shy, and subdued in appearance and behavior. Consequently, many pineland reptiles and amphibians exist below our threshold of perception, at best being recognized as merely a "garden snake" or generic "toad-frog." Species of at least six genera of southeastern snakes: *Diadophis*, *Storeria*, *Opheodrys*, *Thamnophis*, *Virginia*, and *Carphophis*, are typically relegated to "garden snake" status, a meaningless epithet applied to any small and, from the common perspective at least, nondescript snake. As long as this lack of knowledge persists, an appreciation for the conservation issues facing rare species such as *Tantilla oolitica* and *Diadophis punctatus acricus* will be difficult to impart to the general public.

Consider the Pine Woods Snake, perhaps the most overlooked and oft forgotten pine habitat endemic serpent. Because I am fascinated by circumstances that cause some species to fall through the cracks of familiarity, I like to get a sense of which species are widely recognizable to fellow herpetologists, so whenever I have field companions and find something a little different from the typical catch, I show it to my colleagues and ask them to identify it. Among the snake species occurring in the eastern United States, none is more likely to stump the herpetologically savvy than *Rhadinaea*. I have found more than a dozen of these delicate snakes through the years, all but once while I was in the company of other snake enthusiasts, and the only colleague who recognized the Pine Woods Snake immediately was one who, like me, had directed his focus toward pineland herpetofauna. Yet, from the naturalist's perspective, *Rhadinaea* is just as amazing a creation of these forests as the everywhere infamous Eastern Diamondback Rattlesnake.

Description

A small, slender snake that on close examination is seen to be attractively colored. The overall coloration is a rich, reddish brown, very similar to the color of the dead pine straw that carpets the floor of its habitat. Occasional specimens sport more earthen hues, ranging from medium to dark brown. On most adults, faint suggestions of a dark vertebral stripe can be discerned under good light. The top of the head is distinctly darker than the body and exhibits a nebulous pattern of smudged marks. A dark band runs from the nare to the angle of the jaw, stretches across the margin of the upper labials, and passes through the eye. The labials, below the dark band, are set off sharply from the surrounding hues by being yellow, or occasionally white, and an alternate name for the species is Yellow-lipped Snake. In specimens from Florida and Georgia, the yellow coloration is usually clean and immaculate, while in snakes from Mississippi and, to a lesser extent, those from the Carolinas, the light labial coloration is muted by a suffusion of dark stippling and mottling. Allen (1939) reported that the tail is fragile, and many adult snakes are found with tails stubbed to varying extent (plate 5.7).

Adults attain a maximum length of approximately 390 mm (Mount 1975). No detailed description of the size, morphology, or color pattern of hatchlings has been published. Ashton and Ashton (1981) stated that juveniles of an unspecified age class do not notably differ from adults.

Distribution Notes

The Pine Woods Snake's distribution is limited to the Atlantic and Gulf coastal plains and most of Florida. It occurs south at least as far as Palm

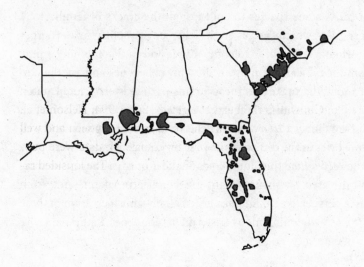

Beach County, Florida (Gottlieb 1984; Reichling and Louton 1989) and may extend somewhat farther but is apparently absent in the Everglades. *Rhadinaea* is a taxon with the potential to inhabit the hammocks of the Upper Keys and pine rocklands in Dade County and the Lower Keys, in disjunct populations similar to *Diadophis*, but has yet to be discovered there.

Northward, *Rhadinaea* inhabits the pine woods of coastal Georgia, South Carolina, and North Carolina, reaching its northern terminus in Beaufort County (Beane 1990). Elsewhere than Florida the distribution lies close to the coast, and penetrates westward from the eastern seaboard only so far as Aiken County in South Carolina (Young 1988) and does not contact the Fall Line.

Rhadinaea has been taken at localities along the Florida panhandle (Palis 1992; Enge 1994) though it is not known if the distribution is continuous along the Gulf coast. Moving farther west, one finds locality records for Mobile County, Alabama (Nelson et al. 1994), and Harrison County, Mississippi (author's observation).

The western terminus of *R. flavilata's* distribution lies somewhere in Louisiana's Florida Parishes. It is likely that the species does not extend as far into Louisiana at the present time as it historically did, and indications are that it has been largely extirpated from the state. Extensive ecosystem damage has taken place in the western longleaf pine belt, endangering the survival of the Black Pine Snake, Gopher Tortoise, Dusky Gopher Frog, Eastern Diamondback Rattlesnake, and Pine Woods Snake in this region.

The obscurity of the lifestyle of this taxon, even within herpetological circles, is amply demonstrated by the paucity of natural history information in the literature, and, paradoxically, the abundant distributional records that are published. On first consideration it would seem that a species for which, in one herpetological journal alone, there appeared nine independent collection and locality coordinate notes (Young 1988; Reichling and Louton 1989; Beane 1990; Godwin 1992; Irwin et al. 1993; Enge 1994; Nelson et al. 1994; Wood 1998; Turner 2003), is a species both commonly seen and well studied. However, during this same 15-year time span, no data were published regarding *Rhadinaea* ecology, reproduction, or diet. The lopsided ratio of geographic notes to natural history notes indicates a species for which simply finding a specimen is noteworthy. At the same time, such locality records yield no insight into the behaviors and lifestyle of a species. For that type of knowledge, significant numbers of wild specimens, or meticulous observation of a few captives, is required, and neither of these resources has been at hand to the small number of herpetologists who might be interested in such investigations.

Perhaps owing to a time now lost, when much larger parcels of undisturbed pine woods existed, there have been instances when it was possible to collect a sufficient sample of Pine Woods Snakes to gain valuable insights. Allen (1939), the seminal Florida herpetologist who collected the seldom-seen Short-tailed Snake with relative ease, reported collecting 21 Pine Woods Snakes in a single day, and boasted that they could be found "at any time" in some locations. This feat may have been something only Ross Allen could do, and only early in the twentieth century.

Habitat

Most Pine Woods Snakes are discovered under objects or beneath loose bark on fallen logs. These are easy refuges to inspect and thus a biased picture of the preferred microhabitat of this species may have developed. A long term sampling of small, secretive herpetofauna would be the kind of effort expected to reveal abundant insights into the lifestyle of a snake such as *R. flavilata*. Few such studies have been attempted to date, and fewer still have been successful in securing a significant number of Pine Woods Snakes from which to derive information.

Based then exclusively on the experiences of field collectors, it would seem that one important component of good Pine Woods Snake habitat is an abundance of dead pine trees, standing or fallen, to provide hiding places for the snake and its prey. Some pine savannas are nearly devoid of such features, such as young successional stands lacking old timber, or sites that experience unusually frequent or intense fires that destroy dead timber or render it unsuitable for small snakes to hide in by burning away the bark covering (plate 5.8).

Another essential ingredient for good habitat is the presence of mesic microenvironments. *Rhadinaea* is usually found in moister settings than is typical of other small snakes like *Stilosoma*, *Tantilla*, and *Diadophis*. As described previously, savanna habitat encompasses a wider range of variations than flatwoods or ridge designations, some being noticeably wetter than others. The proximity of wet flatwoods, which often abut savannas, is similarly a factor in predicting the presence and abundance of the Pine Woods Snake. Campbell and Stickel (1939) reviewed the literature concerning this taxon up to that time, citing microhabitats under logs, piles of straw, and bark, always in settings with abundant moisture such as marshy grounds or ridges that transected swamps. They also commented on a report of one specimen escaping capture by fleeing into water. Their personal experiences with the Pine Woods Snake confirmed its tendency to avoid xeric habitat, as they noted specimens encountered in moist locations embedded within landscapes generally offering dry, sandy situations.

Ecology and Natural History

Myers (1967) stated that 2–4 eggs were laid during the summer, but provided no details. Allen (1939) reported a clutch consisting of 3 eggs measuring approximately 23 x 8 mm. Incubation period was brief, as Allen indicated, no more than 40 days, but whether this was an anomaly cannot be determined without more samples. Hatchlings average 167 mm in total length, of which the tail comprises one fourth.

Like the Scarlet Kingsnake, *Lampropeltis triangulum elapsoides*, which is commonly found in pine woods, *Rhadinaea* seems prone to secreting itself under the bark of standing or prone dead pine trunks during cold weather, and is most frequently found in this setting during late winter and early spring. Unlike other secretive species such as the Florida Crowned Snakes and the Key Ringneck Snake, the Pine Woods Snake is rarely found prowling out in the open, for example, during rainy nights. Invariably, specimens are discovered hiding beneath some object. Where savanna interfaces with more mesic habitat, the burrows of crayfish are utilized by *Rhadinaea* (NatureServe 2005).

The Pine Woods Snake feeds primarily on small frogs and toads, but also upon lizards such as *Anolis*, *Eumeces*, and *Scincella*. Mount (1975) reported that captives avidly fed on Cricket Frogs, *Acris* spp., and Neill (1954) observed captives taking Oak Toads, in addition to anoles and cricket frogs. Newly hatched *Rhadinaea* would find few anurans sized within their capacity to overpower, and so their feeding ecology may differ from the adults.

The Pine Woods Snake overpowers its prey with the help of mildly venomous saliva. Duvernoy's salivary glands secrete a mixture of compounds from the base of the rear upper teeth. The two most posteriorly situated maxillary teeth are slightly enlarged and separated from anterior-lying teeth by a gap (Neill 1954). *Rhadinaea* grasp their prey in their mouths and begin to vigorously chew, bringing these enlarged rear teeth into contact with the prey. After the maxillary teeth have penetrated the prey, the snake becomes motionless for up to three hours while the venom slowly incapacitates the animal (Neill 1954). Small size, docile nature, and venom delivery system renders *R. flavilata* incapable of envenomating a human.

Conservation

Although infrequently seen throughout a patchy distribution of disjunct enclaves, the Pine Woods Snake is probably not threatened except in regions where wholesale destruction of habitat is occurring. The snake appears to be reasonably well adapted to surviving in altered habitat, such as vacant lots on the edges of cities where abundant trash cover provides shelter and supports suitable prey (Reichling and Louton 1989). Other detritophilic forest snakes such as *Tantilla* and *Diadophis* continue to inhabit pine tracts after years of fire suppression, so it is reasonable to suggest that *Rhadinaea* is relatively insensitive to variations in pine forest structure caused by local fire management practice. The taxon is not commonly found roaming and thus road mortality has not been identified as a significant factor in its wild status, further favoring it to coexist near concentrations of human activity. The Pine Woods Snake is of no interest to reptile collectors or hobbyists, being neither gaudy nor valuable. Although direct supporting evidence is lacking, it is possible that the Red Imported Fire Ant, *Solenopsis invicta*, poses a threat to populations in some regions (see Southern Hognose Snake, *Heterodon simus*, species account). Other than profound degradation or complete elimination of pine savanna habitat, the overall conservation status of *R. flavilata* appears to be a secure one.

Summary

The Pine Woods Snake is a species unlikely to become well understood in the foreseeable future. Most literature accounts center on two aspects: the chance collection details of single specimens and behavioral observations of occasional captives. An intimate understanding of *R. flavilata* will require the long-term study of wild populations through sampling of many individuals. These sorts of investigations are the bread and butter of snake ecologists. Most U.S. snake taxa have been served by at least one comprehensive autoecological study by a dedicated student of the taxon, and have had sizeable portions of their natural histories illuminated. Work continues to advance on these species as each new generation of herpetologists yields an individual who picks up where the last author left off, or fills gaps neglected by a predecessor, and gradually a clear picture of a lifestyle emerges.

No such research pedigree has been established for the native species of *Rhadinaea*, although a rich literature exists for the numerous Latin American species, led by the work of Larry David Wilson. The chronic avoidance of delving deep into study of the habits of the Pine Woods Snake is understandable given the experience of Whiteman et al. (1995), who sampled herpetofauna continuously for 23 years before finding *Rhadinaea*! Their 30-year work was situated in South Carolina pineland habitat and yielded more than 6,000 snakes. During the final seven-year period of sampling, nine *Rhadinaea* were captured, prompting the single most data-rich distributional note ever published regarding this taxon.

Tan Racer
Coluber constrictor etheridgei

Perhaps no other pine woods snake is less appreciated for its unique qualities than the Tan Racer. The 11 geographic variants of American racers share numerous unifying characteristics, including a striking ontogenetic pattern transformation, unusually varied diet, spirited defense when captured, and, of course, movement like streak lightning. Another trait they all share—save one—is the propensity to thrive in a variety of habitats. In most rural parts of the southern United States, racers can be seen, and often quite conspicuously. In shaded forests and sunny fields, near swamps and atop dry ridges, while on a backcountry hike or mowing the lawn, you may encounter a racer.

But although racers are a familiar sight to naturalists living in the south, what is commonly not appreciated is that the Louisiana-Texas pine savanna variant is more than an oddly colored version of a common snake.

The Tan Racer is the most habitat specialized among its congeners. A dramatic illustration of the inextricable association of Tan Racers and longleaf savanna can be seen if one stops to examine roadkill snakes during springtime drives in rural southwestern Louisiana and extreme eastern Texas. Richard Etheridge first made this point in a letter to Larry David Wilson while the latter was a graduate student studying the racers of Louisiana and east Texas. Etheridge described an 80.5 km drive spent picking up racers killed by cars and noting the transition from the Eastern Yellowbelly Racer, *Coluber constrictor flaviventris*, to the Buttermilk Racer, *C. c. anthicus* and finally, *C. c. etheridgei*. Within this short stretch of Texas road, uniformly light blue snakes (*flaviventris*) give way to increasingly more spotted specimens, followed by snakes showing a gradual posterior encroachment by tan into the bluish ground color until the blue pigment is limited to a small patch on the nape. Another stretch of road where this rapid transition can be observed is Louisiana Highway 27 running south through Beauregard, Calcasieu, and Cameron parishes. On both of these routes, the entry into longleaf pine savanna corresponds almost precisely with the onset of Tan Racers. The Tan Racer's singular dependency on a specific habitat renders it the only member of its genus for which continued survival is at risk.

Description

The base color is pale grayish tan, upon which are scattered a multitude of tan flecks, barely lighter than the ground color. Tan Racers look as if their scales are afflicted with some dermatological condition akin to toenail fungus, but no two specimens are exactly the same with respect to the amount of spotting. Most adults are abundantly marked, but occasional snakes sport only a few scattered single light scales. The extent of spotting is at least partly ontogenetic, with a steady increase in spotting each time a snake sheds its skin. The venter is immaculate pearly white. Immature Tan Racers, like other congenerics, look dramatically different than adults. The juvenile pattern is marked by a series of dark reddish brown middorsal blotches becoming more faded posteriorly, with smaller dark flecks along each side (plates 5.9, 5.10, 5.11).

Because of the small distribution and the sympatry of two similar subspecies (*flaviventris* to the south and *anthicus* to the east, west, and north), one doesn't have to travel far from the geographic center of distribution to encounter snakes showing the coloration influence of intergradation. In Beauregard Parish, for example, just south of De Ridder, the *Coluber* are a dark bronze color, quite different from typical *C. c. anthicus* but not true *etheridgei* either. Wilson (1970) described intergrades between *anthicus* and *etheridgei* from Texas as having dark bluish patches on their napes, spotted with light pigment similar to the coloration of *anthicus*, but otherwise colored like *etheridgei*. In Louisiana the race is most abundant and phenotypically pure around the vicinity of the town of Fields. In eastern Texas, near the interface of *etheridgei* and *anthicus* habitats, spotted racers have a ground color intermediate between the blue greens and tan that define the two subspecies in their pure form.

Distribution Notes

The Tan Racer has been recorded in Beauregard, Vernon, and Calcasieu parishes in extreme southwestern Louisiana, and Jasper, Polk, Orange, Hardin, and Tyler counties in east Texas. At the eastern margin of its range, Tan Racers appear rather abruptly near habitat where *C. c. anthicus* occur. Just north of De Ridder, in Beauregard Parish, one encounters Buttermilk Racers. In the environs immediately outside De Ridder, specimens similar to the *C. c. etheridgei* phenotype are found, but they are a darker chocolate brown color. Immediately south of the town, along Highway 27, the racers are typical *C. c. etheridgei*. The transition in coloration from one subspecies to another takes place over a distance of no more than 8 km. The change in phenotype corresponds to a noticeable transition from relatively well-

drained high ground in the north to a region of low, wet flatwoods farther south.

Wilson delineated the subspecies in the region and discerned affinities based on color and pattern, after finding no reliable morphological differentiations. In his analysis, Wilson (1970: 67–85) concluded, "The relationship of the spotted forms is reasonably clear. *C. c. etheridgei* would appear to have evolved from *C. c. anthicus* by the development of a uniform tan coloration. . . . The closest relationship of the spotted forms would seem to be with *C. c. flaviventris*, because of the resemblance in basic ground color."

Habitat

The mesic longleaf forests of Beauregard and Calcaseiu parishes are a good place to appreciate the distinction between ridges and savannas. Only a short distance away, in Vernon Parish, is classic pine ridge, and the soils there are extremely sandy and well drained. Surface water in ridge areas is restricted to the intervening drainage bottoms and is ephemeral even there. In contrast, the substrate in prime Tan Racer habitat is often damp, and there are low areas interspersed throughout that stay muddy most of the year. In fact, it is not unusual to find bog associations where pitcher plants dominate the scene. Soils are darker and richer here as well. This is low country and drainage is poor, unlike Louisiana and east Texas pine ridges. Because this habitat lacks the hallmark saw palmetto, one hesitates to call it typical flatwoods, but the conditions certainly represent a transitional state between savanna and flatwoods-type environments (plate 5.12).

Ecology and Natural History

The few direct observations made suggest that in seasonality and clutch size, the Tan Racer does not differ from better-known regional conspecifics. Tennant (1985) described the experience of the Houston Zoo with a wild-caught gravid female that laid a clutch of 30 eggs on May 28.

Discovering snake eggs in situ is a rare event for even the most common species. A clutch of *C. c. anthicus* was found on a sunny bank at the side of a gravel road in Kisatchie National Forest in Louisiana. Like all racer eggs, these were immediately identifiable as such by their rough, scaly texture, reminiscent of the coarse salt sprinkled on a pretzel. They were loosely gathered and nonadherent to each other. These 9 eggs were buried only a few centimeters under the surface in sandy loam. The Tan Racer eggs laid at the Houston Zoo were like these eggs in overall appearance.

A common thread reflected in descriptions of pine woods herpetofauna abundance is the caveat that no matter how uncommon they appear to be, most are probably more numerous than they seem because of their secretive or outright fossorial nature. The Tan Racer is the opposite of this characterization. Far from being inconspicuous, this bold snake makes its presence obvious. In fact, one of the best ways of seeing Tan Racers is simply to drive back and forth across 3 or 4 km of paved road transecting pine savanna habitat within their range. On any sunny day, from spring through fall but particularly during April and May, Tan Racers can usually be seen either crossing the road or basking alongside it, or killed by passing cars. Exploring the forest on foot in these locations is more laborious and far less effective because of the snakes' rapid and almost soundless flight into the tall grasses as soon as it senses an approacher.

To observe Tan Racers in situ it is advantageous to time the hunt to coincide with the first hour in the day when direct sunlight reaches the forest floor, when the racers begin to prowl but are still comparatively sluggish. Overturning pieces of plywood and corrugated metal discarded at dumping sites, which are commonly seen alongside tire-tracked sand roads leading off paved roads in this area, is productive on cloudy or rainy days, when the snakes remain inactive under shelter, as well as during the winter months in this region of mild winters. When fully active and on the prowl in their tall grassy habitat, Tan Racers are unapproachable and nearly impossible to catch.

Tan Racers seem to be adapted to blend cryptically into the light hues of grasses and sand in their habitat. Not only are the light tans perfectly suited to match the color of dead bluestem grass from which the snake is never far, but the lighter splotches break up the outline of the body and make the snake hard to see when it's not moving. Juveniles differ from consubspecific cohorts by being more reddish in color, perhaps also an adaptation that helps them blend inconspicuously into the pine straw carpeting the forest floor.

Tan Racers are the first large snakes in their range dependably encountered each spring, and they are often observed active and abundant on sunny afternoons in March. Based on this early activity and the timing of oviposition, it can be inferred that breeding takes place from late March through early April.

If caught, racers excrete a pungent musk, which makes them less palatable to some animals. Wilson (1970) noted missing tail tips on the majority of Louisiana *Coluber* he examined, including *C. c. etheridgei*. Few snakes are able to purposefully break their tails as a defensive measure, and none are able to regenerate missing tails, as many lizards are able to do, so the explanation for the high percentage of missing tails noted above is uncertain.

Racers defend themselves vigorously when captured, and are the most likely endemic pine woods snake to bite when handled. Bites are delivered with real commitment and very effective in causing a person to drop the snake and lose it. Roger Conant (1958) described a method for catching snakes that is perfectly suited for dealing with the speed and pugnacity of Tan Racers. He recommended grabbing the snake by the tail as it crawled away and quickly flinging it between the legs, clamping together tightly with the knees so the snake is controlled. Then the collector should slowly pull the snake out until the head can be safely seized just before it comes free. This is a spectacular technique for racers. Their speed means that a tail is usually all that can be grabbed anyway, and even that requires alertness, quick reflexes, and coordination. The moment a racer is caught in this fashion it will be trying to swing its body up and bite, so quickly clamping the snake between the legs effectively negates this attack. I have captured many whipsnakes and racers over the years that would have otherwise escaped were it not for Conant's neat trick.

Conservation

Within proper habitat, the Tan Racer is still abundant and conspicuous. Proper habitat, unfortunately, is becoming scarcer with each passing year. As recently as the early 1990s, a cursory stroll near Fields in Beauregard Parish on a sunny April morning would be rewarded with the encounter of several specimens. Recently, most of these fine savannas were auctioned off and sold to five separate interests, so the tracts are no longer managed cohesively, and some have been cut bare and left open. At the present time the subspecies is not as commonly seen, and the general character of the habitat more closely resembles the open fields broken by occasional small forest patches that the Buttermilk Racer regularly inhabits in the parishes to the north.

Concern for this taxa stems from the very limited geographic expanse where good pine savanna habitat persists, and the sympatry of similar but more generalist adapted conspecifics. With the Tan Racer's fitness to pine savanna habitat as the only apparent reproductive isolating mechanism, the fracture and degradation of this forest type could quickly lead to the loss of this odd-looking form. Intergradation with adjacent *Coluber constrictor* subspecies, which has been apparent since the subspecies description, is more widespread now as pinelands have been rapidly cleared and converted to pasture. More so than in other pine-dominated areas west of the Mississippi, the forests, once cut, are often put to use for grazing livestock rather than being replanted with crop pines. This practice increases available habitat that *C. c. anthicus* and *flaviventris* are able to exploit, and steadily shrinks the range of *etheridgei*.

Summary

The Tan Racer is a species strangely ignored by herpetologists. Despite this snake's conspicuous and diurnal behavior, its existence was overlooked by biologists for many decades, and missed during faunal surveys (Guidry 1953). Only one scientific publication has addressed it in any detail. This was Wilson's taxonomic description, which he augmented with intriguing comments regarding its distribution and status. Some attributes of the subspecies mentioned by Wilson should have drawn ardent ecological research for years to come. In the ensuing decades, the taxon has received only one thoughtful treatment (Tennant 1985). The Tan Racer still stands as the most interesting member of its species. With its distribution shrinking and the purity of phenotypic adaptations being obscured by intergradations, a golden opportunity to study the ways in which snakes are shaped by fitness constraints of the southern pine woods may already have been lost.

Plate 1.1. The ground surface of many pine woods is fully exposed to the sun and parched most of the time. Bienville Parish, Louisiana.

Plate 1.2. A relatively closed canopy with mature pine trees is indicative of pine forest structure. Natchitoches Parish, Louisiana.

Plate 1.3. An open canopy and sparse, stunted vegetation denote a pine scrub structure. Bienville Parish, Louisiana.

Plate 1.4. A seasonal wetland in pine habitat. Baldwin County, Alabama.

Plate 1.5. Pine flatwoods. Hardin County, Texas.

Plate 1.6. Abundant palmetto palms often indicate the presence of damp environments in pine woods. Baldwin County, Alabama.

Plate 1.7. Pine savanna with an understory dominated by ferns. Decatur County, Georgia.

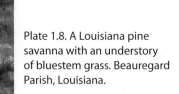

Plate 1.8. A Louisiana pine savanna with an understory of bluestem grass. Beauregard Parish, Louisiana.

Plate 1.9. Pine ridge. Harrison County, Mississippi.

Plate 1.10. The high and dry character of ridge habitat is clearly imparted by this view. Harrison County, Mississippi.

Plate 1.11. Coastal strand pine scrub. Baldwin County, Alabama.

Plate 1.12. Coastal strand pine woods are prime areas for beachside development. Baldwin County, Alabama.

Plate 1.13. Pine rockland. Big Pine Key, Florida.

Plate 1.14. Many pine rocklands have very sparse ground vegetation and extensive exposure of the limestone bedrock. Big Pine Key, Florida.

Plate 1.15. Solution holes are among the few semipermanent sources of freshwater in pine rockland. No Name Key, Florida.

Plate 1.16. Even in mature stands of pine, the actions of wind and water keep most canopies very open in pine rocklands. Big Pine Key, Florida.

Plate 1.17. Pine rockland is aesthetically pleasing for residential development and relatively easy to prepare. Big Pine Key, Florida.

Plate 1.18. Under some common management regimens, the impact on wildlife habitat is severe. Bienville Parish, Louisiana.

Plate 1.19. With sensitive management, commercial silviculture operations can coexist with many types of wildlife habitat. Bienville Parish, Louisiana.

Plate 1.20. Pine rockland. Big Pine Key, Florida.

Plate 1.21. Pine flatwoods. Apalachicola National Forest, Florida.

Plate 1.22. Slow-moving rivers, sloughs, and estuaries create abundant wetland habitats in the Apalachicola lowlands of Florida.

Plate 1.23. High pine ridge in the trans-Mississippi sandhills of central Louisiana. Natchitoches Parish, Louisiana.

Plate 2.1. An abrupt juxtaposition of residential development and wildlife refuge in the pine woods. Baldwin County, Alabama.

Plate 2.2. A delicate layer of fallen leaves and vegetation carpets the ground in many tracts of pristine pine woods. Franklin County, Florida.

Plate 2.3. A fallen log, hacked to pieces, is the consequence of thoughtless reptile hunters. Bienville Parish, Louisiana.

Plate 2.4. A Gopher Tortoise burrow. Harrison County, Mississippi.

Plate 2.5. Pushed-up mounds created by a tunneling pocket gopher. Bienville Parish, Louisiana.

Plate 2.6. A stump hole is a favored refuge for Pine Snakes, Eastern Diamondback Rattlesnakes, and Corn Snakes. Stone County, Mississippi.

Plate 2.7. A pile of discarded lumber can be a gold mine to a field biologist. Jackson Parish, Louisiana.

Plate 2.8. Corrugated roofing tin holds warmth and moisture against the ground and creates a haven for small reptiles. Winn Parish, Louisiana.

Plate 2.9. A borrow pit near a housing construction site offers some amphibian species good breeding habitat. Baldwin County, Alabama.

Plate 2.10. Transitions between forests and clearings can be abrupt in commercial pine stands, creating habitats that are beneficial for some species and avoided by others. Bienville Parish, Louisiana.

Plate 3.1. Louisiana Slimy Salamander, *Plethodon kisatchie*. Origin unspecified. Photo credit: Suzanne L. Collins, The Center for North American Herpetology.

Plate 3.2. Mesic forest habitat of *Plethodon kisatchie*. Fallen logs provide an ideal microhabitat. Natchitoches Parish, Louisiana.

Plate 3.3. Pine Woods Treefrog, *Hyla femoralis*. Origin unspecified. Photo credit: Barry Mansell.

Plate 3.4. Breeding site for *Hyla femoralis*. Baldwin County, Alabama.

Left: Plate 3.5. Oak Toad, *Anaxyrus quercicus*. Lee County, Florida. Photo credit: Christopher E. Baker.

Below: Plate 3.6. Slowinski's Corn Snake, *Pantherophis slowinskii*. Bienville Parish, Louisiana.

Plate 4.1. Flatwoods Salamander, *Ambystoma cingulatum*. Origin unspecified. Photo credit: Barry Mansell.

Plate 4.2. Mabee's Salamander, *Ambystoma mabeei*. Origin unspecified. Photo credit: Barry Mansell.

Plate 4.3. Vernal pool where *Ambystoma mabeei* breeds. Berkeley County, South Carolina.

Plate 4.4. Gulf Coast Box Turtle, *Terrapene carolina major*. Male exhibiting white head markings. Gulf County, Florida. Photo credit: Suzanne L. Collins, The Center for North American Herpetology.

Plate 4.5. Gulf Coast Box Turtle, *Terrapene carolina major*. Female exhibiting reduced carapace markings. Gulf County, Florida.

Plate 4.6. Habitat of *Terrapene carolina major*. St. Vincent Island National Wildlife Refuge, Gulf County, Florida.

Plate 4.7. Brownchin Racer, *Coluber constrictor helvigularis*. Origin unspecified. Photo credit: Suzanne L. Collins, The Center for North American Herpetology.

Plate 4.8. Habitat of *Coluber constrictor helvigularis*. Gulf County, Florida.

Plate 4.9. Apalachicola Lowland Kingsnake, *Lampropeltis getula meansi*. Patternless phase. Captive bred. Photo credit: © 2005 Bob Johnson, ReptileArtistry.com.

Plate 4.10. Habitat of *Lampropeltis getula meansi*. Liberty County, Florida.

Plate 5.1. Dusky Gopher Frog, *Lithobates sevosus*. Harrison County, Mississippi. Photo credit: Christopher E. Baker.

Plate 5.2. Larva of *Lithobates sevosus*. This specimen is diseased. Harrison County, Mississippi.

Plate 5.3. Glen's Pond breeding site and general habitat of *Lithobates sevosus*. Harrison County, Mississippi.

Plate 5.4. Post-burn in De Soto National Forest. Harrison County, Mississippi.

Plate 5.5. Glen's Pond during its annual dry phase.

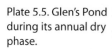

Plate 5.6. Mimic Glass Lizard, *Ophisaurus mimicus*. Origin unspecified. Photo credit: Barry Mansell.

Plate 5.7. Pine Woods Snake, *Rhadinaea flavilata*. Origin unspecified. Photo credit: Barry Mansell.

Plate 5.8. Habitat of *Rhadinaea flavilata*. Pinellas County, Florida.

Plate 5.9. Tan Racer, *Coluber constrictor etheridgei*. Tyler County, Texas. Photo credit: Troy J. Hibbitts.

Plate 5.10. Tan Racer, *Coluber constrictor etheridgei*. Unspotted phase. Beauregard Parish, Louisiana.

Plate 5.11. Tan Racer, *Coluber constrictor etheridgei*. Juvenile. Tyler County, Texas. Photo credit: Terry Hibbitts.

Plate 5.12. Habitat of *Coluber constrictor etheridgei*. Calcasieu Parish, Louisiana.

Plate 6.1. Striped Newt, *Notophthalmus perstriatus*. Marion County, Florida. Photo credit: Christopher E. Baker.

Plate 6.2. Breeding pond utilized by *Notophthalmus perstriatus*. Location unspecified.

Plate 6.3. Pine Barrens Treefrog, *Hyla andersonii*. Origin unspecified. Photo credit: Barry Mansell.

Plate 6.4. Florida Worm Lizard, *Rhineura floridana*. Origin unspecified. Photo credit: Barry Mansell.

Plate 6.5. Florida Worm Lizard, *Rhineura floridana*. Close-up view of head. Origin unspecified. Photo credit: Barry Mansell.

Plate 6.6. Bluetail Mole Skink, *Eumeces egregius lividus*. Origin unspecified. Photo credit: Barry Mansell.

Plate 6.7. Habitat of *Eumeces egregius* ssp. Location unspecified.

Plate 6.8. Southern Hognose Snake, *Heterodon simus*. Origin unspecified. Photo credit: Barry Mansell.

Plate 6.9. Southern Hognose Snake, *Heterodon simus*. Juvenile. Origin unspecified. Photo credit: Barry Mansell.

Plate 6.10. Short-tailed Snake, *Stilosoma extenuatum*. Silvery phase. Origin unspecified. Photo credit: Barry Mansell.

Plate 6.11. Short-tailed Snake, *Stilosoma extenuatum*. Tan phase. Origin unspecified. Photo credit: Barry Mansell.

Plate 6.12. Black Pine Snake, *Pituophis melanoleucus lodingi*. Mobile County, Alabama.

Plate 6.13. Southern Pine Snake, *Pituophis melanoleucus mugitus*. Captive bred.

Plate 6.14. Fire-suppressed pine woods within the range of *Pituophis melanoleucus*, but no longer supporting populations. Rapides Parish, Louisiana.

Plate 6.15. Louisiana Pine Snake, *Pituophis ruthveni*. Bienville Parish, Louisiana.

Plate 6.16. Louisiana Pine Snake, *Pituophis ruthveni*. Close-up view of head. Bienville Parish, Louisiana. Photo credit: jack-kenner.com.

Plate 6.17. Hatchling *Pituophis ruthveni*, the largest neonate among North American snakes. Captive bred.

Plate 6.18. *Pituophis ruthveni* spends the majority of its time underground, as in this specimen photographed in situ.

Plate 6.19. Peninsula Crowned Snake, *Tantilla relicta relicta*. Origin unspecified. Photo credit: Barry Mansell.

Plate 7.1. Key Ringneck Snake, *Diadophis punctatus acricus*. Big Pine Key, Florida. Photo credit: Natalie R. Ehrig.

Plate 7.2. Rockland habitat of *Diadophis punctatus acricus*. No Name Key, Florida.

Plate 7.3. Type locality of *Tantilla oolitica*. Miami, Florida. Photo credit: Steve Conners.

Plate 7.4. Rim Rock Crowned Snake, *Tantilla oolitica*. Origin unspecified. Photo credit: Barry Mansell.

Plate 7.5. Micro-habitat of *Tantilla oolitica*. Dade County, Florida.

Plate 7.6. A prescribed burn in *Tantilla oolitica* habitat. Dade County, Florida.

Plate 8.1. Ornate Chorus Frog, *Pseudacris ornata*. Origin unspecified. Photo credit: Suzanne L. Collins, The Center for North American Herpetology.

Plate 8.2. Ephemeral pond utilized by *Pseudacris ornata*. Baldwin County, Alabama.

Plate 8.3. Gopher Tortoise, *Gopherus polyphemus*. Origin unspecified. Photo credit: Suzanne L. Collins, The Center for North American Herpetology.

Plate 8.4. Eastern Glass Lizard, *Ophisaurus ventralis*. Origin unspecified. Photo credit: Suzanne L. Collins, The Center for North American Herpetology.

Plate 8.5. Six-lined Racerunner, *Apidoscelis sexlineata*. Origin unspecified. Photo credit: Suzanne L. Collins, The Center for North American Herpetology.

Plate 8.6. Southern Fence Lizard, *Sceloporus undulatus undulatus*. Origin unspecified. Photo credit: Suzanne L. Collins, The Center for North American Herpetology.

Plate 8.7. Eastern Indigo Snake, *Drymarchon corais couperi*. Origin unspecified. Photo credit: Suzanne L. Collins, The Center for North American Herpetology.

Plate 8.8. Summer wetland habitat of *Drymarchon corais couperi* in the western extreme of its range. Harrison County, Mississippi.

Plate 8.9. Eastern Coachwhip, *Masticophis flagellum flagellum*. Dark phase typical of populations in the western portion of the distribution. Bienville Parish, Louisiana.

Plate 8.10. Successional meadow in an old clear-cut, with forest ecotone visible in the distance. Bienville Parish, Louisiana.

Plate 8.11. Eastern Hognose Snake, *Heterodon platirhinos*. Natchitoches Parish, Louisiana.

Plate 8.12. Scarlet Kingsnake, *Lampropeltis triangulum elapsoides*. Origin unspecified. Photo credit: Suzanne L. Collins, The Center for North American Herpetology.

Plate 8.13. Eastern Diamondback Rattlesnake, *Crotalus adamanteus*. Lake County, Florida.

Plate 8.14. Habitat of *Crotalus adamanteus* after a prescribed burn. The beneficial effects will soon be apparent. Harrison County, Mississippi.

6

Ridge Specialists

Striped Newt
Notophthalmus perstriatus

It was a golden opportunity to preserve a precious wildlife haven. The site was well known to biologists. They had in fact been discussing how best to protect this rare gem of pine woods habitat. For years conservationists had been debating and wrangling over the best ways to insulate the biological treasure trove from the fervent bustle of commercial silviculture that surrounded the site. Although the value of the land for biodiversity preservation was far broader than the protection of a single species, one taxon had always dominated the discussions.

Tucked away in a remote corner of a vast industrial pine forest was a small depression. The basin spanned the area of a two-car garage and was surrounded by a scrubby mix of longleaf and sand pine, xerophyllic dwarfed oaks, and patchy clumps of palmetto, yielding abruptly to rows of loblolly pine laid out like corn rows. With some variation in timing from year to year, the depression overflowed with accumulated rainfall each spring, spilling the excess out across the adjacent woodland and creating short-lived wetlands. The excess water served to cleanse the water collected in the basin, and the short duration of the hydroperiod, never exceeding six months, guaranteed that no fish were in residence. Living within the square kilometer around this ephemeral pond was a population of Striped Newts, *Notophthalmus perstriatus*, the only ones remaining in this corner of the state.

For several years, stakeholders in the issue of the best use of the land surrounding the pond had held annual meetings with the intent of formulating a compromise strategy. The political climate had changed in the early years of these meetings. No longer could the conservation interests find support from government to raise the state listing of the Striped Newt to Endangered and thus force the establishment of a protected reserve around the pond—certainly not in the face of certain protest from the timber companies that owned the land. Neither could the timber companies afford to ignore the calls of conservation biologists to tread lightly near the pond and exclude the immediate area from any modification that more efficient pine agriculture might demand. Promoting an ethic emphasizing environmental sensitivity was now a public relations necessity for any corporation involved in resource extraction from the pine woods. With these dynamics in play, both sides had been implementing a tenuous compromise whereby the pond and an area of 2 km^2 surrounding it were exempted from logging and winter burns were being applied at the expense of the timber company. This resulted in some reduction in profit for the company as a result of reduced acreage for harvest and the considerable cost of preparing for, conducting, and controlling prescribed fires. Conservation biologists were far from happy about the situation because the area set aside from commercial activities was only marginally large enough to maintain the Striped Newt population, and silviculture remained intense and destructive to natural vegetation immediately outside the border of the sanctuary. On the other hand, industry representatives correctly pointed out that the degree of compromise and the burden of cost they were bearing, all for activities that lessened rather than increased profit margins, were greater on this patch of land than in any other throughout their holdings. So, while the company touted their environmental stewardship and conservationists pushed for more concessions, the newts remained under close scrutiny by a team of scientists who monitored the amphibians, keeping hope that the population would remain stable until an opportunity to protect a much greater swath of land appeared.

Then, abruptly, the entire situation took a dramatic turn. The economic forces that drove the fortunes of the U.S. timber industry had been gradually shifting against it for several years. Many companies had been responding to the rising cost of production and the shrinking monetary return for their products by shortening harvest schedules in order to wring more dollars from their fixed acreage. Eventually, the increasingly attenuated stand rotations dictated that the crop being harvested was too young to produce anything other than pulp for the manufacture of paper products. This was a trend fought against by the conservation camp and had been a key point of contention between the two factions during discussions over the management of the Striped Newt breeding site. Finally, the timber company announced that it could no longer maintain profitability and had decided to sell the land.

The news set off a flurry of activity within the network of those who had been championing the management of the breeding pond site as a natural area. During the years of meetings between the stakeholders, a level of care to maintain the characteristics of good newt habitat had been established, and the amphibians had survived. However, only procedures that did not significantly impact a profitable pine harvest had been agreed upon, so Striped Newt conservation had always been a second priority—a stepchild dependent on hand-me-down actions of no threat to silviculture. Now, it seemed, there was a once-in-a-lifetime chance to secure this tract as a permanent preserve where no compromises would stand in the way of the reclamation of this land as a seasonal pine flatwood wetland. Perhaps now, the conservationists hoped, flatwood wetland conservation would come first.

It would take more than the availability of the land to secure it as a natural preserve. It would require a great deal of money. The land was to be sold, not given away, and this land was valuable. Despite the presence of the newt pond and a small area nearby that was subject to seasonal flooding, most of the overall tract was high, dry savanna and ridge country. These sandy uplands remained accessible to silviculture and logging machinery the entire year, even during periods of heavy rainfall that turned other timber-planted tracts into muddy quagmires and put operations at a standstill. Thus, to keep raw materials flowing continuously to the mills and avoid disruption of product supply and employment of the workforce, the land containing the Striped Newt habitat made the impending sale very attractive to competing timber companies operating in the region. The owners were more than happy to find a buyer who would manage the land as a natural area, but they were going to have to make a competitive bid for the property. Unfortunately, the conservation interests found they were unable to find sufficient funds to purchase the land outright. All the years of talk, interest, and cooperation that the stakeholder meetings had seemed to generate proved to lack the strong foundation of commitment necessary to allocate sizable funds for purchasing the land. All the conservation entities, both public and private, admitted that other funding priorities prevented them from purchasing the site.

As options for securing the newt pond dropped away one by one, those wishing to transform the site into a natural management area began to recognize the futility of their campaign and became resigned to losing the fight. The entire 8,000 ha tract was put on the public auction block with no restriction on how the land could be used. Two forest products companies that held land in the region and had more profitably transitioned to pulpwood production a decade earlier vied for and ultimately bought up the land in a divided patchwork fashion. These companies believed that with their efficient pulpwood mill operations, the chance to increase their base acreage would only improve their financial health.

The company whose land included the newt pond noted the responsible practices of its predecessor and pledged to continue that policy. However, with the consortium of stakeholders now dissolved, the spotlight on the issue diminished and there was no longer any outside consultation to temper management decisions affecting the pond. Whether by conscious design or because of unintentional drifting from the tenets of good environmental management wrought by the stakeholder group, the pond and its surrounding landscape began to slowly congeal into the tree farm that enclosed it. Winter burns became less frequent, then were suspended altogether on account of the buildup of ground fuel that made even low-intensity fires a risky undertaking. The intense silviculture near the environs surrounding the pond continued, and each season a little more of the forest buffer was nibbled away and turned over to that purpose. As the intense site preparation of churning and chopping the soil crept closer to the newts' summer habitat, individuals were killed and those that were not were constrained into an ever-smaller space. The company began broadcasting herbicides after a routine operations management meeting determined that competitive vegetation was becoming a problem for a recently planted sector near the pond, and these chemicals soon found their way into the pond during heavy rains. Adult newts continued to migrate each winter, but fewer recruits emerged from the tainted water and larval mortality increased dramatically.

There was once a golden opportunity to preserve critical habitat for rare pine woods amphibians, but the opportunity was lost.

Description

A complex life history results in five distinct forms. The aquatic form of adults is overall dark olive green, although occasional specimens may be darker, almost brown in base color. The dark green hue covers all dorsal and lateral surfaces of head, body, and tail, but stops abruptly near the venter, where a sharply demarcated yellow color begins. On most specimens, a middorsal lightening imparts a faintly visible vertebral stripe. The undersurface is entirely dull yellow, sometimes marked with scattered black spots. The diagnostic feature of *N. perstriatus* is the pair of unbroken red stripes along the dorsolateral flanks. In all other *Notophthalmus* species, similar markings are broken, reduced to a row of spots, or absent entirely. These hallmark stripes are usually broken at their termini on the head and base of the tail, but continuous on the body. The flanks, between the stripes and the yellow ventral color, often contain tiny red spots (Bishop 1941; Mecham 1967). Befitting their aquatic existence, these adults exhibit lateral compression of the tail, which helps them propel themselves through the water, and indeed they are capable of darting away from threats at amazing speed. This is a smaller species than the Eastern Newt, *N. viridescens*, with adults averaging 52–59 mm in total length, 27–41 mm SVL (Mecham 1967).

During the period of time when sexually mature newts live away from the aquatic environment, they take on a different morphology. In these terrestrial adults, the caudal fin is lost, and the skin becomes slightly rougher. Color and pattern remain the same as when living in water (plate 6.1).

Paedomorphic adults possess external gills and are the only stage that may lack the identifying dorsolateral stripes, with specimens frequently displaying broken stripes or spots along their flanks. As these gilled individuals are essentially sexually mature larvae morphs, their tendency to display markings similar to the pattern of the putative basal taxon *N. viridescens* is an example of phylogenetic reiteration during penultimate ontological stasis.

The terrestrial immature efts are patterned like adults, with the exception of a dull red or orange ground color, rather than green, that presumably is an aposomatic advertisement of their unpalatability. They differ markedly from other life stages by their rough, pebbly skin texture and lack of any trace of lateral tail compression, both characters befitting their habitation of dry environments.

Upon hatching, the larvae, approximately 8 mm in total length (Mecham and Hellman 1952), are marked quite unlike the other four life stages. Two broad, dark bands, located where the red stripes will appear, are the most prominent markings. Postocular stripes are also evident. Red markings are initially absent, and become visible when a larva approaches 30 mm SVL

(Petranka 1998). As in the paedomorphic stage, the red markings resemble those of *N. viridescens* subspecies more than those they will acquire as typical adults.

Distribution Notes

The Striped Newt occurs over a large portion of southeastern Georgia and the north central Florida peninsula, eschewing locations near the coast, and in a second, smaller area encompassing the Apalachicola lowlands and into southwestern Georgia. However, as a result of the specific and rare nature of its habitat requirements, the actual presence of *N. perstriatus* within these broadly defined regions is very spotty, and this is yet another pine woods species that appears more widespread and abundant from distribution maps than in fact it is.

In Georgia, the Striped Newt occurs in scattered areas of suitable habitat throughout the southeastern quadrant of the state, reaching its northern terminus in Richmond County. Its western limit of occurrence in Georgia is uncertain but is likely near the vicinity of the Spring River drainage in Miller or eastern Early counties (as inferred from Johnson 2003). Johnson (2005) indicated that only two significant metapopulations are known to occur in Georgia, one in the Joseph Jones Ecological Research Center and the other at Fort Stewart Military Installation.

A gap in recent records exists in Florida, in the area between the eastern border of Apalachicola National Forest and the Suwannee River. There are scattered pockets of habitat that appear suitable for *Notophthalmus* in

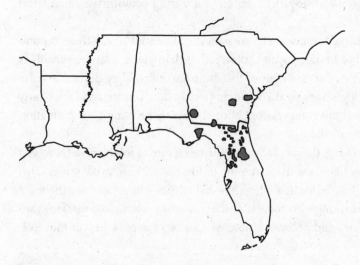

this region, and it may be here that the two putative disjunct enclaves are connected or were during recent times. Within the known distribution of *N. perstriatus* in Florida, actual populations are widely scattered and some consist of single demes centered around one breeding site. The true strongholds of the Striped Newt in Florida are the metapopulations located within Apalachicola National Forest and Ocala National Forest, Jennings State Forest, the Katherine Ordway Preserve, and Camp Blanding Training Site.

The distribution pattern of *N. perstriatus* provokes contemplation. The species' span of occurrence does not resemble that of any other pine woods reptile or amphibian. The larger of the two disjunct enclaves is reminiscent of those taxa—*Stilosoma*, *Rhineura*, and *Tantilla relicta*—that evolved into distinctive forms during isolation on the interglacial islands that preceded the formation of the Florida peninsula. The modern ranges of these insular relicts are restricted to Florida or, in the case of *Rhineura*, make only the slightest incursion across the border with Georgia. In contrast, *N. perstriatus* occupies extensive territory in southeastern Georgia, casting doubt on a possible genesis through isolation on paleo-Floridian islands. Many pine woods reptiles and amphibians occupy territory where *N. perstriatus* is found, but all of these are also known to extend into pinelands in South Carolina and southern Alabama, at least historically, and most are also known to have recently occurred farther north and west. The Striped Newt, as far as documented specimens reveal, has never been found outside northern Florida and southern Georgia, making the limits of its range unique and constrained by unknown factors. Whether this zoogeographic enigma merely indicates the existence of additional undiscovered populations, an unnoticed recent contraction in distribution span, a much older, paleological event, or ecological traits peculiar to this taxon remains open to speculation.

Habitat

Whereas some pine woods species are tolerant of the gradual change in forest structure that occurs during extended periods of fire suppression, the Striped Newt is completely dependent on fire to maintain suitable terrestrial microenvironments. Greenberg et al. (2003) found that among 18 amphibian species monitored in the vicinity of ephemeral ponds embedded in longleaf uplands, only *N. perstriatus* was more abundant in well-burned savanna than in fire-suppressed sites exhibiting hardwood encroachment. These authors postulated that some pine woods amphibians are tolerant of floristic changes caused by fire suppression, including the Oak Toad and the Pine Woods Treefrog, among endemic pineland species. In contrast to these generalists, the number of terrestrial adult newts captured was significantly less compared with trapping success in forests experiencing regular 2–3-year fire cycles.

Upland habitat utilized by newts during their terrestrial phases is typically very open, with sparse shade cover. Wiregrass is the dominant understory plant, and the lushness of the understory is reduced compared with adjacent locations where newts are less abundant. Leaf litter represents an important type of surface cover for newts, but it is often sparsely distributed. Stumps and stump holes represent other types of refuge that compose significant portions of suitable microhabitat at some sites.

The physical structure of breeding sites and larval habitat also reflects the species preference for regularly burned patches. Wiregrass remains the dominant plant growing within these shallow basins. Aquatic light levels are high, quite distinct from some of the situations where Oak Toads and Pine Woods Treefrogs breed. The combination of xeric pine scrub experiencing routine fires and topographic and hydrologic features that facilitate the dependable formation of temporary ponds has likely always been a rare habitat, and the Striped Newt is naturally a species occupying a patchy distribution. The modern-day forest management trend of supplanting fire with mechanical thinning and herbicide treatment, along with residential development of high pine areas, has pushed the Striped Newt toward a more precarious status in the wild (plate 6.2).

Ecology and Natural History

Among broad habitat categories, the southern pine woods are the harshest and most limiting environments in the southeastern United States for herpetofauna. Amphibians, tied as they are to moist microhabitats and, in most taxa, annual availability of persistent surface water reservoirs, are particularly challenged in the pine woods. Amphibians survive through modifications of morphology, physiology, and ethology, driven by unique natural selective forces. Any successful amphibian design in this difficult setting must be able to bridge long periods when critical resources become very scarce or disappear. They must also possess the faculties to quickly exploit brief explosions in resource availability that may arrive unexpectedly in local areas. No other pine woods amphibian displays more plasticity in its life history traits than *Notophthalmus perstriatus*, and thus none is more responsive to the peculiar vagaries of this landscape.

Striped Newts breed in ephemeral pools that form in winter or early spring. The larvae require a minimum of two to three months of standing water to complete metamorphosis. Even when this basic requirement is met, the specific timing of pool availability and duration can vary widely from year to year. Striped Newts are able to take advantage of sudden changes in their surroundings that can be exploited for reproduction. They are resilient in the face of short hydroperiods and well equipped to maximize durable optimal conditions during good years.

Ephemeral ponds are created by closely spaced days of drenching rains. The migration of adult newts from their scattered refugia in the dry sandhills to basins is triggered by these weather events (Dodd 1993). Some wet periods sufficient to create standing pools of water are not adequate to fill dry bowls with enough water to persist very long. Striped Newts respond by making multiple immigrations and emigrations to and from the ponds until enough water accumulates in the catchments to provide a reasonably safe environment for eggs and larval development. These rainfall-triggered migrations to breeding pools can vary temporally, occurring any time between December and March, even between consecutive years (Bishop 1941).

The duration of the larval period could not be more dynamic. This is an essential trait for an amphibian with an obligatory gilled larval stage living in an environment that capriciously supplies aquatic habitats on an unpredictable schedule. Considering this stochastic resource availability, it is amazing that any amphibian is able to occupy such sites as anything other than a transient presence. Yet the Striped Newt not only survives under such conditions, it is endemic to areas that inflict these kinds of challenges, and its life history is a reflection of the consequent struggle.

One of the most critical survival strategies of *N. perstriatus* is plasticity in the length of its larval period. Once larvae reach a threshold size where metamorphosis to the terrestrially adapted eft stage is possible, the newts have two sharply contrasting options. If a pond begins to dry early, forecasting the approach of conditions unsuitable for gilled amphibians, the larvae will rapidly transform into efts and leave the pond to spend one to two years in the xeric uplands while they complete their sexual maturation (Dodd and LaClaire 1995). In populations where local conditions supply dependable, persistent winter or springtime ponds, all or a subset of a larval cohort may retain most features of larval morphology but continue to mature and attain sexual maturity in this state. These paedomorphic newts remain in the pond for an extended period and breed in this condition (Dodd and Charest 1998; Johnson 2002). However, unlike some other salamanders in which paedomorphosis occurs, *N. perstriatus* paedomorphs are obligated to transform into terrestrial adults after reproducing and thus able to survive a slow pond-drying event (Johnson 2002, 2003). This spectacular flexibility in a life stage that is typically the most vulnerable for amphibians allows the Striped Newt to escape deadly conditions as they frequently arise in the pine woods setting, and yet also permits it to make the most of times of plenty.

Striped Newts are generalist predators and exhibit the seasonal shifts in proportion of various prey species typical of an opportunist. The species feeds throughout the year. Christman and Franz (1973) dissected 59 *N. perstriatus* from a depression marsh in Alachua County, Florida, and found invertebrates composed the overwhelming number of taxa among identifiable remains. However, it should be recognized that chitenous insect fragments are resistant to digestion and thus more readily discerned than, for example, small amphibian eggs and larvae. Coleopteran, trichopteran, and dipteran larvae were frequently consumed by Christman and Franz's sample group. A temporal subset of specimens that had fed largely on lepidopteran caterpillars was attributed to flooding during the weeks surrounding sampling that was presumed to have washed normally inaccessible prey into the water. During other periods, newts had fed almost exclusively on the eggs of Spring Peepers, *Pseudacris crucifer*, mirroring the well-documented proclivity of the congeneric *N. viridescens* to prey on the eggs of various frogs and toads. Mecham and Hellman (1952) also noted *N. perstriatus* to be an important predator of *P. crucifer* eggs.

Conservation

The preservation of any pine woods species with a restricted distribution is best accomplished by properly managing expansive swaths of habitat. Ideally, the protected land should be larger than the minimal area required by the resident population, thus providing a buffer against activities beyond the preserve boundary and affording areas for population expansion. It should be possible, however, to save relict populations of many species by assuring the land actually occupied by the majority of the population is protected.

Not so the Striped Newt. Its complex life history and the harsh environment in which it lives require a broader landscape approach to conservation initiatives. Semlitsch (2002) defined the issue of amphibian conservation as one requiring a good understanding of the total land occupancy of a particular species during the course of a year. Often, attempts to protect a rare amphibian species are focused on the delicate breeding sites that, due to their comparatively tiny size, are more vulnerable to complete annihilation than the surrounding habitat. However, Semlitsch pointed out that just as important is the protection of an area surrounding these sites of a size sufficient to accommodate the normal migrations to and from the ponds. These terrestrial preserves need to be maintained in accordance with the natural history of the species of concern, with proper microhabitats, prey base, and corridors to the breeding sites. With the exception of the Louisiana Slimy Salamander, which moves but little during the course of a year, all the endemic pine woods amphibians are as dependent on expansive tracts of natural flatwoods, savannas, or ridges as they are on transient aquatic features.

Further complicating the conservation of the Striped Newt is its predisposition for local extinction. The harsh nature of its habitat, in particular the natural scarcity of reliable springtime pools in the forest, results in a tenuous presence at any one location under the best of circumstances. The taxon persists over a broader landscape through metapopulation dynamics that are possible only when corridors of natural habitat connect demes. As one site becomes unable to provide suitable habitat for breeding or larval development, some individuals will move to other wetlands. In its current state, the habitat of the Striped Newt throughout its global distribution is ardently maintained in optimal condition at a number of good tracts, but these are isolated from other areas by poor habitat that represents movement barriers. The local populations are secure only as long as the environment remains stable, but the vagaries of climate and the constant threat of disease, such as the crisis affecting the Dusky Gopher Frog, make the current conservation strategy very unstable. One of the few realistic ways to address this vulnerability is through captive breeding and the translocation of propagules to reclaimed sites.

Summary

The Striped Newt is the only pine woods reptile or amphibian that can be wholly aquatic its entire lifetime. Yet this distinction describes only a subset of individuals, and the newt also displays the most complex and variable life history of any member of the endemic herpetofauna. The newt is so variable

in its environmental needs, from locale to locale, and its population numbers so plastic from season to season, that no simple recipe for conserving it
is possible. Only by setting aside very large tracts of habitat that encompass
the full range of habitats used by *Notophthalmus*, from seasonally flooded
areas to xeric ridges, can a viable population of newts be protected for the
long term. Such an expensive undertaking will not be made on behalf of
a species with such a minor impact on the consciousness of the general
public. Conservation is often a public relations and marketing issue, and
only species that become popular owing to their appeal to humans and ac-
cessibility to our emotions and empathy will succeed in garnering much
support. Consider the strong efforts made on behalf of the manatee, which
is marketed very effectively to appeal to the general public, and how difficult
the same thing would be in the case of the Striped Newt.

Pine Barrens Treefrog
Hyla andersonii

The two men in Atlanta's Hartsfield International Airport were not catching
the usual connecting flights for tourists from Switzerland. Instead of heading for one of America's larger cities, these men were bound for Raleigh,
North Carolina. From Raleigh, they would drive a rental car to a rural area
adjacent to a private wildlife refuge in Moore County. At their hotel they
would be joined by a local man, a biologist of some repute, who was essential to the success of their trip.

Their baggage was also rather odd. Scattered throughout the clothing and
toiletry supplies, in a purposeful effort to make them appear less significant,
were flashlights, batteries, hiking boots, insect repellent, and boxes of zip-
lock sandwich bags. On this day the men were lucky, and nothing in their
suitcases aroused the suspicions of Customs officials or airline security. No,
these were not typical tourists, they were wildlife poachers, and their quarry
was one of the most beautiful amphibians in the United States, the Pine
Barrens Treefrog, *Hyla andersonii*, a species protected by the wildlife laws
of every state where it occurs.

Throughout the world there extends a network of reptile and amphibian hobbyists who depart from the principles of the vast majority of such
individuals. They operate outside the laws protecting rare species and are
actually drawn to them above all others. Not that they are unconcerned that

some reptiles and amphibians are being pushed toward vanishing in the wild, but they feel that their skill in keeping and breeding these animals in captivity justifies removing them from the wild, regardless of laws prohibiting or restricting this. Certainly, too, there is a strong profit motive that further attracts some. Although they may deny it, the fact that these species are rare, often extraordinarily beautiful and distinctive owing to evolving in very specific and restricted habitats, and highly coveted by collectors, combines to make the possession of these animals irresistible and worth considerable risk.

While attending a herpetological conference the previous year, the men had made the acquaintance of the biologist from North Carolina. The three struck up a friendship and soon discovered they shared similar viewpoints regarding collecting and breeding rare reptiles and the legitimacy of protective regulations. From this camaraderie was born a business partnership that promised risk, excitement, considerable profit, and, sadly, the elevation of the status of the men in the eyes of a small subculture within the herpetocultural community. They made a quick excursion to the sandhills of North Carolina and scouted for good locations for a return trip the following spring, when conditions would be right for finding their quarry. On the flight back to Switzerland, the plan to return was hashed out in detail, and the two friends talked and thought of little else over the ensuing months.

Every year, poachers remove endangered reptiles and amphibians from their protected habitats in private reserves and national parks. Law enforcement officials are vigilant about these crimes, but an unknown number of perpetrators surely evade apprehension. There are simply too many ways to easily obtain endangered animals in their habitats, and too many unscrupulous people willing to exploit these situations, to catch any but a tiny proportion of criminals. The strong profit motive of the collectors and the covetous desires of a subculture of hobbyists provide a powerful dynamic that maintains a steady trade in illegal herpetofauna despite the best efforts of conservationists.

By the eve of their return trip to North Carolina, the Swiss dealers had been paid cash deposits for 50 Pine Barrens Treefrogs by a dozen amphibian and reptile hobbyists scattered throughout Europe. Many knew little about the wildlife laws of the United States, much less those governing North Carolina, but sensing a crisis of conscience in the offing, they chose not to research the topic too deeply lest they be confronted with a decision to knowingly proceed with an illegal acquisition. Others simply didn't care and freely expressed their disdain for wildlife laws as a matter of pride and superior purpose, declaring that the captive breeding of endangered species by those who truly appreciated their worth and beauty, and the commerce of the offspring to like-minded individuals, was conservation work of the highest calling.

Arriving at the hotel by early evening, the men quickly set about their preparations. They had only four nights to accomplish their goal, and a long soaking rain that was just letting up as they pulled into town promised immediate success. A knock at the door announced the arrival of their American friend, and soon they were all off in his truck headed for the outskirts of town and a patch of low pine woods.

They could hear the din of raucous honks of a breeding chorus as soon as they stepped out of the vehicle. Flashlights were checked and handfuls of plastic bags were stuffed into back pockets. The trip had been timed perfectly to coincide with a breeding congregation. With frogs at such high density and the location so private, the job would be conducted swiftly and safely. Taking one last cautious glance down the road for any approaching vehicles and seeing none, the poachers began making their way down the slippery, grassy slope to the bog below. Upon reaching the soggy border, each man started off alone, carefully playing his flashlight beam along the water's edge and around areas of emergent vegetation farther into the bog. There was little point in trying to track down a specific animal by following his call because the sheer number of males created such a confusing cacophony of sound that this was inefficient. It would be far quicker to search randomly with the flashlight and collect frogs as they were encountered; their numbers were great.

The poachers were abruptly startled by the sound of static punctuated by intermittent beeps, coming from the top of the hill. A moment later, flickering light emanating through the underbrush told of people approaching with flashlights, and the crackling sound of walkie-talkie conversations confirmed their worst fear. Their truck, pulled off onto the side of this remote country road, had caught the attention of a sheriff's squad car as it made its pass down this route. No normal explanation could be made for pulling a vehicle off to the side of this lonely road in the pine forest on a rainy spring night, so the two officers stopped to investigate. Fortunately for the poachers, no frogs had yet been caught, and the only visible gear they carried, flashlights, did not necessarily condemn them as poachers.

The officers carefully approached the men, lights trained directly at their eyes, and asked them what they were doing. The local man spoke for the group and responded with a half-truth in a calm and confident tone. He stated that they were all herpetologists. "That's a scientist who studies reptiles and amphibians," he offered helpfully, "with a professional specialty in frogs." This odd and unexpected confession that they were frog experts would, he knew, throw the officers off guard, as they were expecting to find drug-related activity or, possibly, some sort of game poaching. Seeing no guns, the lawmen were perplexed and susceptible to a tall tale delivered with authority. The local man went on to describe how his two Swiss colleagues had desired to observe the breeding aggregations of treefrogs to compare with their studies of European species, and he continued to describe this research in enough spurious detail to bore the officers, further putting him in psychological control of the situation. Because the elements of his alibi were essentially true, and because the sheriff's deputies were taken completely aback by this bizarre yet seemingly benign eccentricity, the ruse served the poachers perfectly. The officers instructed the men to leave, warning them that the land was private property, but otherwise left them alone and hurried back up the hill to escape the steadily increasing drizzle.

Not wishing to press their luck, and realizing they had narrowly escaped a serious run-in with the law, the three poachers quickly followed the officers, got in their truck, and drove back into town, aborting their mission. Close calls such as this were an occasional part of their line of work, and each man had had his share. There were always other animals, other destinations, to visit. Always, there was a customer anxious to make a deposit at the next opportunity to acquire some seemingly unobtainable species and be the only person known to own such a coveted treasure. The following year, the two Swiss collectors were apprehended at Miami International Airport, caught as they tried to smuggle two dozen specimens of a critically endangered Caribbean snake to a wealthy collector. However, the North Carolina frog collector still makes his living poaching and selling endangered reptiles and amphibians of the southern United States, and has made several successful collections of Pine Barrens Treefrogs from the Moore County bog.

Description

Hyla andersonii is a medium-sized treefrog of unmistakable appearance. Unlike other species of *Hyla* native to the South, which are able to change hue and pattern, *andersonii* maintains a stable appearance. Its dorsum is bright lime green and unmarked. Along each flank a conspicuous broad band, ranging in color from lavender to brown, runs from slightly anterior to the eye posteriorly toward the rear limbs. The band is framed along its dorsal side by a thin white or, in some populations, yellow border. Near the groin the lavender marking dissipates into a field of bright yellow spots. The legs and feet are also adorned in lavender markings that are clearly visible when the frog is disturbed from its resting posture. The venter is white. These markings are unlike the color pattern of any other species of treefrog found in the southeastern United States. Newly metamorphosed froglets are somber versions of the adults, with darker green dorsum and darker lateral bands not so distinctly set off from the dorsal coloration (plate 6.3).

Like many other treefrogs, *H. andersonii* exhibits mild sexual dimorphism. Males range from 35 to 38 mm SVL and are smaller than the more squat females, which range from 38 to 44 mm SVL (Noble and Noble 1923).

Significant differences in coloration between the populations in the extreme south (Florida and Alabama) and specimens from the remainder of the range have been noted (Means and Longden 1976). Those from North and South Carolina as well as from the extralimital populations in New Jersey are deep green in dorsal coloration. *Hyla andersonii* from the Florida panhandle and southern Alabama are a lighter lime green. The dark lateral bands on frogs found north of Alabama are distinctly lavender or purple in color, while specimens from the southernmost populations exhibit brown lateral bands. On Florida and Alabama treefrogs, the thin light line separating the dorsal coloration from the lateral bands is bright yellow in most individuals, whereas in other populations this marking is usually white.

Newly hatched larvae, which measure 4.5 mm in total length (Noble and Noble 1923), are uniform pale yellow with faint darker stippling. More mature tadpoles are clearly patterned with a ragged-edged dark brown stripe running down the length of the tail musculature. The superior and inferior tail fins bear no distinctive pattern, but moderate dark stippling is apparent. The head is dull brown.

A seemingly strange comment by Noble and Noble (1923) that *H. andersonii* emits a strong odor, reminiscent of raw peas, is in fact quite accurate. Many amphibians produce strong, often acrid odors when handled roughly, and these may be quite species specific in character. Presumably this phenomenon is associated with and augments the distasteful or noxious skin secretion defenses that are common to amphibians.

Distribution Notes

Hyla andersonii is relegated to a few, very widely scattered enclaves within the southern pine belt. Its distribution is one of the most interesting, and puzzling, of any of the species covered in this book. Certainly the present-day distribution is relict, but the former extent of the species' distribution will never be fully knowable.

In North Carolina, the Pine Barrens Treefrog lives in scattered sites across 16 counties in the south central sandhill country (Bullard 1965).

The South Carolina distribution, nearly 965 km distant from the Florida sites, encompasses eight counties, but the strongholds of the species are situated in Chesterfield and Kershaw counties. As in North Carolina, the range lies adjacent to the Fall Line.

Christman (1970) first reported the presence of *H. andersonii* in Florida. Initially thought to be restricted to boggy locales in Okaloosa County, it was subsequently found over a wider area, encompassing Walton, Santa Rosa, and Holmes counties in addition to Oklaoosa County.

The species is also known from a few sites in Alabama, in Escambia County, as a continuation of the locales just across the Florida state line.

Neill (1948) reported a specimen from Richmond County, Georgia. This remains the only record for the state. The single specimen was found resting beneath the bark of a standing tree, during winter, and this description is thus also noteworthy as one of a very few of the in situ activities of *H. andersonii* at times other than during breeding congregations.

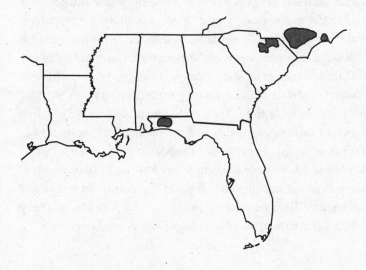

In addition, the Pine Barrens Treefrog has long been known from its namesake habitat in New Jersey. Although paralleling many of the characteristics of the southern pine belt habitats, this ecoregion lies outside the scope of this book.

Habitat

The hyperdispersed populations that dot the overall distribution of this amphibian are centered around small sites presenting dramatic microenvironmental transitions over a small spatial scale. The Pine Barrens Treefrog is best designated a ridge specialist, but this requires some elaboration because the species is actually associated with a very specific set of environmental circumstances that are embedded within ridge habitat. Defining the habitat of *Hyla andersonii* in the context of pine ridge inhabitant requires a further dissection of this type of landscape to understand the components important to this species. The Pine Barrens Treefrog's breeding versus general occupancy habitats contrast sharply, a situation similar to that of the Dusky Gopher Frog. However, in the Carolina sandhills and Florida panhandle pine ridges, the disparate elements are juxtaposed even more abruptly than they are in the range of the Dusky Gopher Frog.

Soligenous bogs form along the lower levels of sandy ridges. Porous substrate along the top of the slopes is necessary for bog formation and the cause of dramatic contrasts in moisture level and vegetational character found in sites supporting Pine Barrens Treefrogs (Sharitz and Gibbons 1982). The quickly draining sandy caps atop ridges allow rainwater to percolate into the substratum with little lost to surface runoff. Some ridges are situated over hard-packed foundations of clay. In other cases, the slopes intersect high water tables at their juncture with the flatter base terrain. In either scenario, water filters through the upper layers of the ridge until it encounters an obstacle that directs its drainage laterally, causing it to weep from the hillside and trickle down the surface of the remaining slope to collect at the bottom and persist for considerable time. The development of characteristic bog vegetation, such as carnivorous plants and the hallmark sphagnum moss (*Sphagnum* spp.), indicates the presence of saturated ground and a hydroperiod sufficiently wet to provide standing water for lengthy periods.

Hyla andersonii breed in soligenous bogs to the near exclusion of any other type of habitat. Their occurrence in a particular location is largely determined by the presence of soligenous bogs (Means and Longden 1976).

The water at boggy breeding sites is acidic. Gosner and Black (1957) measured pH ranging from 3.6 to 5.2 in breeding pools in New Jersey, and stated that average pH of sphagnaceous bog water was 4.0.

A delicate balance must exist among fine-scale physiography, soil characteristics, and annual rainfall level and seasonal pattern to create bogs in scattered locations throughout the Atlantic and Gulf coastal plains. Means and Longden (1976) noted that wide variation in adult microenvironments suggested that *H. andersonii* is not highly specific in its requirements as an adult, but that larval habitat (bogs) were remarkably similar between sites across the entire distribution. They reasoned that the availability of bog habitat was a significant limiting factor in the life history of the species and its presence a good predictor of *H. andersonii* populations at specific locations within generally suitable pine ridge habitat.

Glacial climates in North America are believed to have fostered more widespread and less disjunctly distributed bog habitats during the Pleistocene, and the Pine Barrens Treefrog probably occurred fairly continuously throughout the coastal plains of the southeastern United States in such situations (Means and Longden 1976). Means and Longden further developed their hypothesis by stating that the greatest expanse of Pleistocene bog habitat was located in the milder southernmost clines, where present-day populations reside in Florida and Alabama. New Jersey, they believed, was not as suitable an environment for bog habitat as it has become in recent times. Therefore, they postulated the Florida and perhaps the Carolina populations to be relicts of a long-standing presence, but characterized the New Jersey populations as the comparatively recent descendants of "southern invaders" (Means and Longden 1976).

Ecology and Natural History

Heavy rain showers during the appropriate season stimulate breeding congregations. In the Florida panhandle, breeding choruses may be heard as early as April and will recur periodically at least as late as September (Means and Longden 1976). As would be expected, the breeding season is later in northern enclaves (Noble and Noble 1923).

Most calling males are situated in arboreal perches, partially hidden in the vegetation surrounding the margins of water-filled basins or slowly flowing seepage courses. Noble and Noble (1923) observed one calling male situated 2.4 m above the ground, but heights of 91–121 cm are more typical (Means and Longden 1976). Most calling is concentrated during the early evening, even if the temperature and humidity remain optimal and rain falls later into the night.

The voice of male *H. andersonii* is reminiscent of a quacking duck, and has been variously described as a "quak-quak quak-quak," "aquak-aquak aquak-aquak," or "keck-keck keck-keck" (Noble and Noble 1923). Noble and Noble also reported that Deckert, a Florida herpetologist, described the call as a single syllabular "whang," which surely indicates a misidentification of the species that was vocalizing.

Spawning takes place in slightly moving, acidic spring flows or pools. Amplexus is supra-axillary, and the embracing male fertilizes eggs as the female extrudes them in groups of 1–4. Fertilized eggs fall to the bottom of the pool, where they lie singly over the sphagnum moss and other vegetation. As many as 1,000 eggs may be laid by a female during a single breeding event, although in some populations, females typically lay as few as 200–300 eggs (Means and Longden 1976). Eggs hatch in 3–4 days at normal springtime temperatures.

The larvae spend much of their time lying motionless. Noble and Noble (1923) observed them aggregating in the shallow margins of pools, so shallow in fact that the upper surface of their tails broke the surface of the water as they lay on the bottom. These aggregations are skittish, and they scatter immediately upon the slightest disturbance, retreating to vegetation in deeper sections of the pool.

Hyla andersonii requires approximately one year to complete larval development and metamorphosis. During the protracted breeding season experienced by Florida populations, larvae that hatch from late season spawns in August and September may require the whole of the following warm season to continue their growth and will not become sexually mature until nearly two years after hatching (Means and Longden 1976). The larvae cohabit the bog with numerous predatory species, in sharp contrast to another pine ridge amphibian, the Striped Newt, which breeds exclusively in ephemeral rain-filled pools that lack most typical amphibian larvae predators. Fish and snakes such as water snakes, *Nerodia* spp., and gartersnakes, *Thamnophis* spp., are common in *H. andersonii* breeding sites. Independent attempts to rear larvae in captivity have unanimously concluded that this is very difficult to accomplish. Perhaps as a result of a very narrow and atypical water chemistry to which the larvae have adapted, mortalities in lab-reared tadpoles are high.

Conservation

Like the Dusky Gopher Frog and Striped Newt, the Pine Barrens Treefrog is highly vulnerable to alterations upon its aquatic breeding and larval habitat, and this process appears to have been a significant force in its decline. The pocosin bogs where the Pine Barrens Treefrog congregates to breed are a highly threatened habitat, subjected to a distinct set of threats that are different from those affecting *Lithobates* and *Notophthalmus*.

Roughly three quarters of the pocosin bog sites within the southeastern Atlantic Gulf coastal plain, where all but the Florida panhandle *H. andersonii* population reside, have been eliminated in order to put the land into agriculture or for extraction of natural resources (Sharitz and Gibbons 1982). Much of what remains is unprotected and exists tenuously because of economic factors that have temporarily made utilizing the sites unattractive to the owning interests. For example, many pocosins are embedded within industrial forests and remain undisturbed as a consequence of the higher cost of draining and preparing these lands for pine agriculture compared with sites with better natural drainage. Currently, U.S. timber industries are experiencing a depressed market, but the increasing scarcity of wood and wood products must inevitably make it economically attractive to convert these sites over to commercial silvicuture.

A second threat to seepage bogs that is in remission is the harvest of peat. Currently the demand for gardening soils is the primary incentive for peat harvest, which is a relatively modest draw on natural peat reserves. However, peat can be burned and represents an alternative fuel to coal for the production of electricity. Peat can also be used to produce ethanol (Reed et al. 1981). In the increasingly tight U.S. energy market, these optional fuels place pocosin bogs squarely in the sights of government-sponsored energy development projects and public utility interests.

Peat mining is accomplished by draining pocosins and removing the resource down to where it meets the underlying substrate. After the peat has been removed, the land is often converted to agriculture or silviculture to exploit the disruption of the hydroperiod that formerly made it unsuitable for such enterprises, and critical amphibian breeding sites are destroyed. In the face of such powerful, economically driven priorities and the insatiable demand for more abundant and affordable energy, the Pine Barrens Treefrog, which receives no federal protection to deflect habitat intrusions, cannot possibly garner sufficiently influential proponents to champion protection of its habitat.

Anderson and Moler (1986) stated that female *H. andersonii* are predisposed to breed with the Green Treefrog, *H. cinerea*, because of a preference for the slightly different calls of those males. Under normal circumstances, spatial and temporal differences in reproduction act as isolating mechanisms between these two species. The authors described two treefrogs collected in Oklaloosa County, Florida, that they proved were hybrids, one *H. andersonii* x *H. cinerea* and the other an *H. andersonii* x *H. femoralis* cross. The explanation for these hybrids favored by Anderson and Moler was that a drought that preceded the discovery of these specimens constrained sympatric *Hyla* species both temporally and spatially. In the absence of known postzygotic isolating mechanisms, the presence of these three species, all in breeding condition, at the same time and place, allowed interspecies hybridizations. Although this instance of hybridization appeared to be a natural occurrence and on a very small scale, the authors mentioned the potential for similar phenomena to follow where habitat alterations have removed physiographic barriers and blurred the distinction between isolating niches in the reproductive biology of closely related anurans. They cited several examples of this that have occurred with other taxa. The potential for hybridization between the Pine Barrens Treefrog and other *Hyla* demonstrated by the two Florida specimens is cause for concern. Disruption of habitat, such as slight topographical modifications that eliminate seepages and foster the appearance of more typical transient pond environments, could create a situation where breeding congregations of several hylid species would occur at the same location, leaving only easily disrupted temporal differences to prevent hybridization on a large scale. Once a large number of hybrids entered the environment, they could gradually pollute the gene pool of *H. andersonii* and bring the population to its demise. This process might go undetected for a long time, as Anderson and Moler noted that the *H. andersonii* x *H. cinerea* hybrid was phenotypically indistinguishable from *H. andersonii*.

Summary

The spectacular appearance of the Pine Barrens Treefrog is an important factor in its population status and security. The commercial reptile and amphibian hobby is driven by many of the same factors that influence the collecting and dealing of other commodities. Rarity is an important factor in setting demand, and thus price, but it is not the sole consideration. There are other anurans that are just as geographically restricted and scarce as *H. andersonii*, examples being the Florida Bog Frog, *Rana okaloosae*, the Wyoming Toad, *Bufo baxteri*, and the Dusky Gopher Frog. Yet in spite of their rarity, there is no trade in these species, legal or otherwise, because no demand exists for such humble-looking, earthen-toned species. When the rarity of specimens is coupled with singular and vibrant beauty, a powerful incentive to possess these animals is created.

The peculiar, widely disjunct distribution of *H. andersonii* indicates a relict species that was once distributed more widely. In large part, its present-day rarity is the result of natural processes that have shrunk its habitat and relegated it to a handful of persistent sites. This status presents difficult questions when considering efforts that should be expended to protect the treefrog. In a time when resources for wildlife conservation are stretched very thin and the use of such resources struggles to compete with conflicting demands to utilize habitat for human needs, endangered species and their critical habitat areas must be prioritized for actions. The harsh reality is that all declining species and their habitats will not be rescued. Conservation agencies must decide which land and which organisms will be the focus of their efforts, and which issues will be ignored. The Pine Barrens Treefrog is an organism that disappeared in many areas solely through natural processes. Should such a species be protected? Should *H. andersonii* conservation consume portions of the limited supply of funds, personnel, and political capital at the expense of other endangered pine woods herpetofauna? Species such as the Flatwoods Salamander, in contrast to the Pine Barrens Treefrog, would be thriving over a wide portion of the southeastern coastal plain were it not for the destructive practices of mankind over the last century. Flatwoods Salamanders and Pine Woods Treefrogs compete for the capacity of the same federal, state, and private agencies to protect and manage two very different types of pine woods environments, and these parties must unavoidably choose which taxon will be their primary focus.

Florida Worm Lizard
Rhineura floridana

A consideration of the herpetofauna of the southern pine woods conjures numerous biogeographical questions. What processes led to the disjunction of the pine belt at its western limit, and prevented it from extending farther westward into Texas, giving rise to a separate species of pine snake? How did the ringneck snake establish itself in the lower Florida Keys when none occur on the upper Keys? Are the isolating mechanisms that fostered the evolution of a distinctive race of kingsnake in the Apalachicola region of Florida still in force today, or were they prehistoric features that have vanished? The pattern and composition of reptile and amphibian occurrence in the southern pinelands are ripe for such questions because of the paradox presented by the diversity of endemic taxa in a landscape relatively devoid of modern-day natural barriers. In montane landscapes, such as in southern Arizona or the Appalachian Mountains, the reasons for the evolution of endemism are more obvious. The biogeographical forces that contributed to present-day biodiversity in the pine woods are cryptic by comparison.

The movement of great fragments of crust over the surface of the earth—continental drift, or plate tectonics—is a process attributed with creating interesting worldwide distribution patterns of the planet's biota. Marsupial mammals are present throughout Australia and the Americas but absent in Africa and Asia. Alligators and cryptobranchid salamanders are limited to disparate locales in North America and China (and Japan in the case of the salamander). The horned tarantulas of Africa (*Ceratogyrus*) and South America (*Sphaerobothria*) are strikingly similar in appearance and are believed to share a recent common ancestor. These and numerous other puzzling extant animal distribution patterns are explained by prehistoric connections, conventions, and separations of continental plates. A less familiar distributional anomaly is the presence of two of the four extant families of amphisbaenid worm lizards within three widely separated distributions in North America, while the other two more specious families are broadly distributed throughout South America, much of Africa, and Europe and the Middle East. With the lion's share of species residing in the southern hemisphere, how did four species representing two endemic phylogenies at the family level come to reside in three distant pockets of occurrence in Mexico and, of all places, central Florida? An analysis by evolutionary biologist J. Robert Macey and colleagues (Macey et al. 2004) provides the answer.

Ancient Earth landmasses once consisted of two continents, Laurasia in the Northern Hemisphere and Gondwana in the Southern Hemisphere. Approximately 250–290 million years ago (MYA) the two great continents of Laurasia and Gondwana convened into one supercontinent called Pangaea. By the end of the Paleozoic era, all life on Earth was evolving on this single landmass, without continental isolation by intervening oceans. It was at this time that the ancestral worm lizards evolved. The Laurasian and Gondwanan landmasses began to move apart again approximately 200–225 MYA, throughout the Triassic period, thus beginning the phenomenon of independent continental evolution that has held sway ever since. Today's peninsular Florida contains land that was once part of Laurasia. The most parsimonious explanation for the presence of an amphisbaenid in Florida would be that it is a direct lineage of a basal Laurasian ancestor.

Assuming a Paleozoic origin, amphisbaenid taxa now occupying Laurasian-originated North America should be basal, as they were the first to split away from the remaining forms. Comparing mitochondrial DNA sequences from taxa representing all four families of amphisbaenians, Macey and colleagues concluded that the monotypic family represented by the Florida Worm Lizard and three closely related species of the family Bipedidae living in Mexico are indeed the basal sister taxon to all other forms of worm lizard, worldwide. The genomic sequence of Florida's singular worm lizard, translated by the linguistics of molecular biology, reads like a text describing the nature of the planet Earth eons before the dawn of man.

Contemplation of the humblest of creatures can lead to insights into the biggest questions of biology. Like the Pine Woods Snake, whose meek demeanor and anonymous life style cause it to be overlooked by most persons, the Florida Worm Lizard is in reality a treasure trove of information, belying an existence that is undetected and uninteresting to most. This underappreciated marvel is a link to a distant time in Earth's history, and the esoteric methodology used to dissect its genetic structure sharpens our understanding of the processes that generated the diversity of present-day life on the planet, knowledge that everyone seeks.

Description

Nothing else living in the southern pine woods resembles Florida Worm Lizards. They are legless, nearly colorless, and almost blind. They lack external ear openings. The texture of their skin is unique, for although they are scaled, the individual scales are fused into ridges that encircle the body, making them look like some sort of strange annelid worm.

Worm lizards appear to lack skin pigments; their hue is imparted by the physical qualities of the dermis—its thickness, texture, and consequent intensity of blush from the underlying capillaries. Adults are reddish, pink, or purple depending on the size of the specimen and what section is being viewed. The few juveniles that have been described were more purplish, with diminished reddish and pink hues than adults, but morphologically very similar to older animals (Carr 1949) (plate 6.4).

Worm lizards are not completely blind, but their eyesight is rudimentary. There is considerable variation in the prominence of eyes between specimens. Neill (1951b) examined 14 adults and found 10 to be without visible eyes, but the other four specimens did have eyes to varying degrees, ranging from clearly visible black dots to faintly discernible specks lying just below the skin (plate 6.5).

Adults average 20–25 cm in total length (Gans 1967) and the maximum recorded length is 38 cm (Telford 1955). Hatchlings are quite large in relation to adult dimensions, ranging from 80 to 100 mm (Carr 1949). Two juveniles described by Carr had prominent eyes, and he speculated that the eyes are gradually lost as the lizards reach adulthood, an idea that he acknowledged had earlier been advanced by Garman (1883). However, Neill (1951b) found no correlation between prominence of eyes and body size, and no further studies have commented decisively on this question since the appearance of those two publications.

Distribution Notes

Overall, the distribution of the Florida Worm Lizard corresponds to the central Florida ridge country within the xeric sandy sections of the state. This taxon's distribution encompasses the entire breadth of the Florida peninsula between Cedar Key, Levy County, and the Tampa Bay area along the Gulf side, to the Atlantic coast from just north of Cape Canaveral south to St. Augustine (inferred from Gans 1967). It extends its range farther south

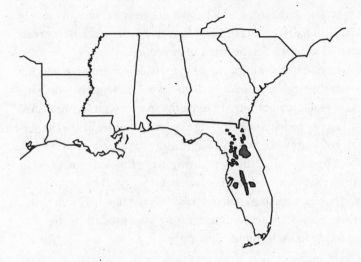

on an inland tract following the central ridges, terminating in the vicinity of De Soto County.

The northern limit of the range is less well understood, but probably occurs in extreme southern Georgia. A single specimen recorded from Georgia is without specific locality details attached, but Meylan (1984) found one specimen in Hamilton County, Florida, only 18 km from the border with Georgia. That portion of Georgia borders the western margin of the Okefenokee Swamp and Suwannee and Suwannoochee river areas, so the most likely habitat for *Rhineura* in Georgia is embedded disjunctly in the upland tracts in extreme southern Echols County.

Habitat

Meylan (1984) described the habitat as longleaf pine/turkey oak scrub atop sandy, friable soils. Neill (1951b) painted a more varied picture of the situations in which *Rhineura* may be found, citing hammocks near water, including those dominated by live oaks, and wiregrass flatwoods in addition to more expected habitat such as turkey oak woodlands and "high pine." From Neill's account it appears that *Rhineura* is not completely restricted to ridge habitat, nor perhaps even to pine-dominated habitat, although the hammocks he referred to may well have been embedded within pine woods. Worm lizards also venture into the barren dune habitat that comprises part of the Lake Wales Ridge ecosystem. However, *R. floridana* is nearly endemic to sites with an overall pine woods landscape, especially ridge habitat. Thus it meets the criterion of inclusion as a habitat endemic because the loss of all of its pineland refuges would result in the taxon existing only in tiny scattered enclaves and teetering on the brink of extinction.

Meylan's observation of the importance of a friable substrate is a weighty assertion, as this character is a good predictor of the presence of worm lizards within its overall distribution. Although the species is an accomplished burrower and not dependent on a yielding substrate to move below the surface, a crumbly soil composed mainly of sand indicates a xeric microhabitat. Given the worm lizard's vulnerability to drowning in flooded ground, well-drained substrates such as sand are a requirement for good habitat. Search for *Rhineura* in habitat similar to that described for the Bluetail Mole Skink, *Eumeces egregius lividus,* by gently finger-sifting sand around the bases of turkey oaks growing in well-drained pine ridges.

Ecology and Natural History

The most fossorial reptile or amphibian in the southern pine woods. This strange creature is totally obligated to spend its entire life several centimeters or more below the surface of loose, well-drained soils. Worm lizards, in contrast to other primarily fossorial herpetofauna of the habitat, are not found after dark roaming roadways or the forest floor, even after rainfall. The only exception to this statement is the phenomenon of specimens, sometimes in considerable numbers and usually deceased, found after flooding deluges, but these animals were fleeing or flushed from saturated subterranean refuges and were not conducting their normal activities. Rarely if ever is *Rhineura* found on the surface beneath objects or in decaying logs (Meylan 1984), again in contrast to other fossorial co-inhabitants of these habitats. Almost all specimens of worm lizards have been found while excavations of some type were being conducted or, as mentioned, after floods have washed them out and drowned them. Consequently, no direct observations of the taxon's activity patterns have been published. As with several other secretive pine wood species (*Ophisaurus mimicus, Tantilla oolitica,* and *Diadophis punctatus acricus,* for example), the entirety of our understanding is feebly based upon chance encounters with single specimens after they have been disturbed.

Entirely as the result of one well-documented discovery, a surprising amount of information is available regarding the taxon's reproductive biology. Carr (1949) unearthed two *Rhineura* eggs in his backyard. He surmised that the neonates were already pipping when he exposed them, but in any event they were escaping the eggshell and quite active by the time he realized his discovery. The exact depth at which the eggs were resting was uncertain, but was estimated to be 20–51 cm below the surface. A thorough search revealed no additional eggs or hatchlings, so the two eggs probably constituted the entire clutch. Gans (1967), citing Neill (1951b), gave the range of clutch size as 1–3, although neither author mentioned examining any specific record of three eggs in a clutch.

From Carr's report we also know that the dimensions of *Rhineura* eggs are unusual. His examples were 8.9 x 38 mm, roughly the proportions of a 1½-inch section of an ink pen. The low fecundity and large hatchling size suggest a reproductive strategy for which success is dependent on high juvenile survivorship.

One might assume that earthworms would comprise the primary diet of *Rhineura*. Captives have thrived on them as their sole diet over extended periods (Neill 1951b). However, the Florida Worm Lizard is not an earthworm specialist. The majority of identified prey remains in the stomachs of museum specimens are from terrestrial spiders, primarily species of wolf spider (Neill 1951b) of the genus *Hogna*. Florida's xeric sandy ridges support abundant populations of specialized species of *Hogna*, which construct burrows and spend almost all of their lives underground in the manner of terrestrial tarantulas. These large, innocuous spiders would be frequently encountered by worm lizards foraging below the surface, and their large size makes them the equivalent meal for a worm lizard as a mouse is for a kingsnake.

Florida Worm Lizards are extremely vulnerable during local flooding events. Their unique morphology and limited mobility render them unable to float or swim even for brief periods. Within their subterranean locations they are overtaken by floodwaters before they have any opportunity to move to higher ground. Meylan (1984) described finding 11 *Rhineura* dead after a major flood of the Suwannee River. The bodies were lying in pools of water and found as the flood was receding. Given the poor dispersal capabilities of *Rhineura*, it would seem that such events would permanently obliterate the taxon from flood-prone zones and cause local extinctions whenever historical flooding of normally dry territory occurs. As rivers change course over time, new flood zones are created and gradually erode the distribution of the Florida Worm Lizard. Although the same process would conceivably create new opportunities for colonization by worm lizards as previously wet areas became more xeric, it is hard to conceive of a means by which they could reach the new territory given their virtually nil dispersal capability.

Conservation

Rhineura floridana feeds heavily on fossorial lycosid spiders (Neill 1951b). This proclivity renders worm lizards vulnerable to habitat alterations subtler than those that observably alter the superficial physical environment. Not only will populations be affected directly by gross scale changes in components such as slope and drainage, which often go hand-in-glove with conversion of ridgeland to agricultural or residential use, but they may also respond to more insidious changes that affect their favored prey, which are themselves adapted to very specific conditions. Any alteration in the sandhill environment that impacts wolf spiders will eventually ripple out and affect local populations of *Rhineura*. Of the species of spiders on which *Rhineura* is most likely to encounter when foraging, a large component are specialized species of the genus *Hogna* that are highly restricted in distribution and specialized to live in the xeric pine ridge sites in central Florida (Edwards 1994). Direct contact with chemical contaminants such as the pesticide runoff from citrus groves in the region might not be directly detrimental to individual worm lizards, but it could certainly have a major negative impact on their prey base. Thus, although chemical contaminants are not usually cited as a directly toxic threat to reptiles, especially those living in environments where aquatic habitats are scarce, a mechanism for significant impact from toxic agricultural run-off and the health of Florida Worm Lizard populations is predicted.

Summary

The Florida Worm Lizard is completely alien to our familiarities. In form and function it resembles nothing else in the southern pine woods. *Rhineura floridana* is the sole representative of a taxonomic order in the United States, the only instance of this among our native herpetofauna, and thus there is nothing physically comparable to it in this country. It is a legless, virtually unpigmented, and nearly sightless evolutionary holdout of an ancient lineage. Even the human response they provoke is unique to them. The worm lizard is a creature so odd and novel that biologists, upon their first encounter of a living example, are able to express little more than childlike marvel.

Worm lizards are the most profoundly subterranean pine woods reptile or amphibian. Consequently, it is the species most hidden from our eyes and most removed from our awareness. Its way of life is largely a mystery and substantial questions abound. What is a typical home range, or does this concept even apply to *Rhineura*? How does it locate prey, or are feeding opportunities merely happenstance encounters? What are its courtship behaviors, and how are mates found at all? The concept of an individual worm lizard living its entire life without ever being exposed to the surface for even a moment is probably accurate for the typical specimen, which is a unique distinction among the pine woods herpetofauna. It is also the reason why most of the basic questions we can ask about it will likely never be fully answered.

Bluetail Mole Skink
Eumeces egregius lividus

It is not the ubiquitous pine trees that provide the critical habitat components that shape southern pine woods reptiles into specialized forms. Casual observers might assume that, because pine trees are visually dominant, they are also the environmental factor most responsible for the endemic forms of wildlife that define the pine woods. If it were possible to surgically excise all the pine trees from the habitat, the reptiles covered by this book could continue their lives without disruption. Of course, such a scenario is a fantasy, because all biotic components of a particular habitat are interconnected in seen and unseen ways, and eliminating the canopy of pines would unleash a cascade of environmental and ecological changes that would eventually lead to grave consequences for all but the most adaptable species. Certainly, the canopy would need to be replaced with a hardwood providing similar shading characteristics. It is also likely that the lack of pine needle ground litter, and its replacement with a deciduous leaf litter, would gradually change the soil and surface water pH, which might exert an effect on amphibians directly, and eventually all fauna by disrupting the food web. But if it were possible to remove the pine trees without disturbing any other environmental feature, it is unlikely that any reptile would be immediately affected. The point is that the interconnection between pine tree and reptile is not direct, and that there are more essential environmental attributes of the southern pine woods than their hallmark trees. Pine trees and pine woods herpetofauna are sympatric species that are both adapted to and dependent on a more fundamental set of characters that are the true definers of the southern pine woods as an ecosystem. To better understand the herpetofauna, it is necessary to identify the basest elements of this environment.

Perhaps the most basic underlying element of the pine woods is its soil, and the Bluetail Mole Skink is a superb illustration of the powerful influence of substrate on the biogeography of a pine woods reptile. All southern pine woods grow on sandy soils. High ridges can be exclusively sand near the surface, while organic buildup in ravines and flatwoods may create a layer of richer ground while still containing a large measure of sand. As any gardener knows, sand helps lighten the soil, making it less inclined to pack and dry rock hard, and thus its presence helps foster microhabitats conducive to small fossorial reptiles.

The five subspecies of *Eumeces egregius* share a predilection for sandy substrates. Although found in a variety of situations, all are closely tied to sandy sites where they can easily secrete themselves below the surface of loose, yielding soil. The wholly pine ridge endemic subspecies, *E. e. lividus*, is the form most closely tied to pine-shaded sand beds. Its marriage to the surface layer of substrate sand is remarkable among its sympatric congenerics and reveals the Bluetail Mole Skink to be a true endemic, shaped by and inextricably tied to the pine woods environment.

Phenotypically similar to the sympatric Southeastern Five-lined Skink, *E. inexpectatus*, only the Bluetail Mole Skink is actually limited by the parameters provided by the central Florida pine ridges. Where xeric uplands transition into other ecological zones, the close relationship between substrate and skink can be appreciated. The species composition of *Eumeces* between abutting pine upland and deciduous woodland reveals an abrupt absence of *E. e. lividus* where the soil becomes more fertile and organic. The entire visual character of these two habitats is starkly different, but it is simply the base quality of sandiness that confines the Bluetail Mole Skink to its areas of occurrence, while holding no such influence over the other similar-looking skink species in the region.

Description

The Bluetail Mole Skink was discovered during a yearlong study of the natural history of a closely related subspecies, *E. e. onocrepis* (Mount 1963, 1965). The vibrant blue tail of *E. e. lividus* is simultaneously the quickest way to distinguish it from conspecifics and the reason it is possible to confuse it with another, more common skink species. None of the other four subspecies of mole skink has a blue tail when adult (Mount 1968), but one sympatric congeneric, *E. inexpectatus*, the Southeastern Five-lined Skink, typically does through subadulthood (Conant and Collins 1991). If a small skink with a bright blue tail is found in central Florida, the most expedient way to distinguish a Bluetail Mole Skink from a Southeastern Five-lined Skink is by examining the stripes on the body. *Eumeces e. lividus* display 2 light dorsolateral stripes, one per side, while *E. inexpectatus* has 5 thin light stripes. Bluetail Mole Skinks have noticeably reduced legs and are of slender build; this combination imparts the appearance of a proportionally longer body than that of five-lined skinks. Other differences involving scalation distinguish mole skinks from other taxa, but for field identification they are not necessary to know.

The body coloration in adults is brown, marked with the distinctive lighter longitudinal stripes. These light brown markings are closer together on the anterior trunk and diverge slightly apart along their posterior extension. The tail of juveniles through young adulthood is brilliant blue. On old specimens a suffusion of pinkish hue begins to supplant the blue; over time this becomes increasingly muted. In some aged individuals the tail is pinker than blue, and such specimens bear resemblance to the parapatric subspecies *E. e. onocrepis* (plate 6.6).

This subspecies may most resemble the ancestral lineage that gave rise to the present-day phenotypic diversity. Mount (1965) believed that the coloration of *E. e. lividus* was the most pleisiomorphic and the one from which other races were derived. As with other herpetofauna restricted to Florida's central ridges, the Bluetail Mole Skink and its ancestral founders have likely occupied these territories continuously since the Pleistocene, whereas other mole skink subspecies inhabit younger land that appeared after a comparatively recent drop in global sea levels.

Distribution Notes

This subspecies is confined to Highlands, Polk, and Osceola counties in central Florida, along the Lake Wales Ridge. The range is discontinuous and fragmented into many disjunct parcels, some quite small. Only 34 locales along the ridge are documented collection sites for *E. e. lividus* (U.S. Fish and Wildlife Service 1999).

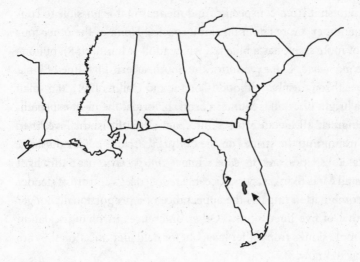

Habitat

Optimal habitat for the skink is patchy. Good sites are dominated by very dry ridge habitat, at elevations >30 m (U.S. Fish and Wildlife Service 1999), with deep, loose sand substrates. The surface of such areas is unvegetated bare sand in many spots, but interspersed throughout are areas where aggregations of xerophyllic oaks and uneven terrain have created a thin layer of leaf litter and moister conditions that support some low-lying vegetation (plate 6.7).

The precise spot favored by Bluetail Mole Skinks is an open area in pine ridge scrub where the sand is so deep and loose it gives the sensation of standing on a beach. The general environment is hot, sunny, and dry. Small piles of sand, excavated by burrowing geotrupinid scarab beetles, are abundantly scattered about, as are the much larger mounds of the Eastern Pocket Gopher, *Geomys pinetis*, and perhaps the entrance to a Gopher Tortoise burrow as well. Only a few yards from this location, vegetation becomes denser, though still sparse, and in barely perceptible depressions a layer of oak leaves and occasional fragments of fallen oak branches are lying. A hike over the general vicinity reveals a heterogeneous patchwork of these juxtaposed bare sandy openings and pockets of dry, vegetated scrub.

Such fine-scale variation in the habitat may be critical for a small reptile unable to escape poor conditions. The skink's need to thermoregulate in open, sunny sites is met by the barren zones. Such hot, dry conditions are unsuitable for egg deposition and for the delicate hatchlings, so the comparatively milder, mesic environment afforded by the partially shaded surface litter among the oaks is probably necessary for mole skink populations to survive.

Ecology and Natural History

One herpetologist, Robert H. Mount, a familiar name to students of south-eastern U.S. herpetofauna, can be credited for virtually all that is known about the natural history of all five subspecies of mole skinks. His exhaustive paper on this species (Mount 1963) is a remarkable body of work. Mount's thorough coverage of most of the facets of the life history of this species leaves no broad topics left to be studied, perhaps in part explaining the paucity of detailed literature on the ecology of this species since 1963, though a few noteworthy exceptions exist. Mount conducted his most detailed studies in Levy County, Florida, outside the distribution of *E. e. lividus*. However, his work took place in xeric pine scrub habitat and most of his observations apply to the endemic pine woods subspecies also. Almost all that follows is a synopsis of Mount's work, so each component is not specifically cited. Where statements are based on or augmented by other studies, these have been cited.

These skinks have an affinity for the loose piles of soil created by the digging of geotrupinid scarab beetles and Gopher Tortoises, and especially the pushed-up mounds of pocket gophers. This is particularly evident in the most xeric habitats, where the majority of specimens are found hiding within such piles. Skinks remain under cover in these mounds, and it is speculated that they move closer to the surface or deeper down in the soil to thermoregulate. Mount stated that mole skinks were most abundant in mounds during winter days under very specific weather conditions. During chilly but sunny days, when the preceding days had not been frigid, Mount found many specimens by digging into pocket gopher and beetle mounds. After lengthy stretches of very cold temperatures, the skinks were not to be found in these situations, as was the case during the summer. Mount reasoned that the loose and dry sandy soil composing the piles absorbed heat more rapidly than the harder packed substrate surrounding them and afforded the best opportunity for the lizards to bask while still remaining hidden from predators. In addition to the characteristic weather conditions needed to draw the skinks to the mounds, the lizards exploited these features only within the open, barren components of their habitat. Even within the right temperature parameters, mounds located in oak thickets or other

shaded situations did not yield skinks when searched. The thermal properties of these mounds that make them such ideal basking sites during mild
winter days also renders them unsuitable in more vegetated habitat. The
reduction in direct sun exposure due to mid and upper story vegetation
prevents the mounds from warming sufficiently beyond the general ambient surface temperatures such that they offer no advantage to the skinks.
In less sun-baked environments, mole skinks eschew the mounds and are
usually discovered under loose leaf litter.

Mount reported that mole skinks in Levy County breed in the fall and
oviposit the following spring, which is not the typical timing of reproductive
events for many reptiles in the Deep South, but is a strategy known for some
taxa such as the Eastern Indigo Snake. The timing of reproduction in *E.
egregius* farther south on the Florida peninsula, including the southernmost
populations of *E. e. lividus*, may differ somewhat from other populations,
as Babbitt (1951) described courtship and mating by a pair of skinks (*E. e.
egregius*) on Key West during March. In most populations of *E. e. lividus*,
copulation takes place during September and October. Males develop bright
orange yellow coloration on the flanks and sides of their head during this
time. Females deposit 2–9 eggs, most often 4 or 5, during April and May
(Mount 1963; Hamilton and Pollack 1958).

As with most members of the genus for which observations have been
made, female Bluetail Mole Skinks brood their clutches. After depositing
eggs several centimeters below the surface, usually under the additional
shelter of some object such as a fallen log, the female forms a circular coil
around the clutch and stays with her eggs for the duration of the incubation
period. The function of brooding by mole skinks is not known. Hamilton
and Pollack (1958) observed a captive mole skink bury her eggs by tunneling
beneath them, causing them to sink into the substrate and become covered
by it, then curling around the eggs and keeping them in a tightly packed
pile. Mount (1963) also observed brooding females and described how they
frequently turned the eggs and licked them. Eggs artificially incubated in
the absence of the female did not hatch if they were removed on the same
day they were laid, but developed and hatched if removed after several days
of female brooding. Mount thought that the licking of the eggs by the females might help prevent spoilage by means of an antibacterial or antifungal
quality in the saliva.

The eggs hatch one to two months after being laid, based on four clutches
observed by Mount. This is an unusually variable incubation period. The

temperature at which these eggs were incubated apparently was not constant among the clutches, as suggested by Mount's statement that temperature explained most of the variation in length of incubation, but no specifics were given. Even with widely disparate temperatures, the range of incubation period is unusually great.

The Bluetail Mole Skink employs the locomotive behavior known as "sand swimming." In this technique, the limbs are held against the side of the body while the lizard moves beneath the surface by vigorously undulating the body in the fashion of a wriggling snake. Species employing this method of underground motion are able to move astonishingly quickly, hence their resemblance to a fish swimming effortlessly through the water. Mole skinks are well adapted to this behavior with their pointed snouts, slender body, polished and tightly lying scalation, and reduced limbs.

Conservation

Referring to the species as a whole, Mount (1963: 359) stated that "within the range of the species, all natural areas of 100 acres or more in extent supporting sandhill or scrub associations are also supporting red-tailed skink populations." While pockets of such habitat still exist in the region, such as in Ocala National Forest, the area specifically inhabited by the Bluetail Mole Skink is not protected by such large swaths of public land. Finding xeric sandhill parcels of 100 acres (40 ha) or more along the central Florida ridges is a depressing task, as very few have survived the intense residential and agricultural development that has taken place over the last 50 or 60 years.

When one considers the portion of the environment actually occupied by the skink, the importance of maintaining the natural character of the first several centimeters of substrate on sites intended to preserve *E. e. lividus* populations becomes apparent. This fragile crust overlaying the pine ridge ecosystem is essential to the mole skink. As a species that neither climbs, basks, or hunts on the surface, nor digs deep burrows, its entire existence takes place in the vertically minute zone of the topmost layer of fallen leaves and pine straw to the point below the sandy surface where pressure compaction makes sand-swimming movements useless, a depth of perhaps no more than 12–15 cm.

Numerous factors may damage this fragile biotic zone. Fire-suppression allows the encroachment of an unnatural diversity and density of understory vegetation that impedes the activities of mole skinks. Paradoxically, fires that are too intense and destroy leaf litter patches and xerophyllic dwarf oak patches will also be detrimental by eliminating mesic refugia for this species. Anything that compacts soil, including obvious things such as off-road vehicles, but also seemingly harmless activities such as foot traffic, can bruise the habitat in ways that are difficult to see. Simply hiking through a sandy ridge may jeopardize the skinks that are living underfoot and unseen.

Summary

Befitting its name, the Bluetail Mole Skink is a species attuned to the attributes of the top several centimeters of substrate. Less crucial to the skink, at least in terms of direct effects, are the qualities of forest structure that may seem to be more obvious and important components of the pine ridge environment. Yet it is not possible to separate one habitat element from another, as all are interconnected in some way and affect each other. Should fires be suppressed for extended periods, woody undergrowth develops, initially creating a dense midstory within the pine forest and eventually overtaking the pines and becoming the dominant upper story component by preventing germination and sapling growth of young pines. With the increase in vegetation and no cleansing fires, leaf litter accumulates on the forest floor. This litter breaks down and forms a carpet of soil upon the surface, obliterating the dry sandy substrate required by mole skinks. So, while the skink is a creature of the soil and does not immediately require the presence of any particular species of pine, or any pine for that matter, the two elements are interconnected. Only fire-adapted pine species provide a fire-resistant overstory that provides some shade and inhibits erosion. Thus only in well-maintained habitat can the loose, sandy substrate remain exposed and intact and provide the appropriate microhabitats for mole skinks.

The mole skink highlights the importance of considering the substrate when managing natural pinelands as wildlife habitat, and the need to respect this feature when visiting such places. Even in generally fine pine ridge scrub, finding areas with pristine ground is difficult. Performing beneficial prescribed burns usually involves many vehicles traversing the forest, and activities such as scraping away the soil to create firebreaks. Throughout the process, the footfalls of numerous persons on fire crews inflict further damage that is rarely noticed or considered. When the task is finished, the forest may appear beautifully natural and renewed, but oftentimes the surface layer of soil has been damaged. Little attention has been directed to the time necessary for a xeric sandhill substrate to recover from such degradations as compaction, the mixing of natural layers, the loss of leaf litter, and damage to surface cover such as the diverse lichens, mosses, and ground-lying vegetation that are characteristic of these areas. These are all important elements of microhabitat to the Bluetail Mole Skink, as they determine the availability and nature of refugia, nesting sites, moisture gradients, and the abundance and biodiversity of invertebrates upon which skinks prey. The lesson to learn from considering the mole skink is that the southern pine woods are far more complex and fragile than they seem.

Southern Hognose Snake
Heterodon simus

My foot was so grossly swollen, it frightened me to look at it. When I stood, it felt dull and almost lifeless, and I could feel the tightness of the skin straining to contain the rapid buildup of fluid. I took a seat in the truck, and pulled down the vanity mirror to look for the explanation for the oddly heavy sensation I was feeling in my face. I received a shock when I saw my reflection. Large clumps of sickly yellow hives hung above and below my eyes. My lips were puffed up like a punch-drunk palooka, and the rest of my face displayed a fiery blush. I was on a remote logging road, a long way from a hospital on this early Sunday morning. Most of my potential rescuers were probably sitting in church, and I was having an anaphylactic allergic reaction to fire ant stings.

A mere 20 minutes before, it had been business as usual around the campsite. One of the trucks had sunk in the sand and I was pushing from the rear while a companion gunned the engine. As I strained against the vehicle, I suddenly felt the pinpoint burns on my ankle that signaled a fire ant attack, a common event out in the pine woods. The enraged soldiers were responding to our audacious gall of disturbing their busy morning with spinning tires. I did the familiar dance while chanting curses as I brushed off a dozen of the little demons. Unknown to me, my many previous encounters with these terrors had gradually sensitized me to their venom.

How appropriate, I thought. Here I was, facing the very real possibility that I would leave these woods as a corpse, and the reason for my being here was to investigate the likely extinction of another frequenter of these pine woods for which these insect invaders were being blamed.

The week before, I was aroused from a slow day at the zoo by a phone call from a young man living near Wiggins, Mississippi. He knew of my interest in the snakes of the region, and believed he had found a Southern Hognose Snake. Earlier that year, I had read a paper by herpetologist Tracy Tuberville and colleagues, based at the Savannah River Ecology Lab, documenting the disappearance of this little reptile over large parts of its former range. The authors hypothesized that the spreading plague of Red Imported Fire Ants was the primary culprit in these extirpations. No confirmed sighting of this snake had been made in Mississippi for decades, so the rediscovery of the species had me packing, and one day later I was in southern Mississippi to check on this possible resurrection and, unknowingly, to face my own brush with extinction due to fire ants.

I drove up to a small brick house situated on the edge of a plowed sandy field and rang the doorbell. A young man, the snake collector, answered and took me around back to a shed, where a rusty oil drum sat. A sheet of plywood weighted down with two cinder blocks sat on top to prevent the escape of any snake that could have possibly fit within. I removed the lid and peered inside. The bottom was covered with straw and no snake was visible. I tipped the drum and began fishing through the straw with my fingers. As I dug down to the bottom, the rough-scaled textures and white, silver, and gray hues of a Southern Hognose Snake came into view, but a suspicion I'd been harboring had kept me ready to react quickly. Sure enough, the snake in the drum was a Pigmy Rattlesnake, not a hognose! An easy mistake to make, I thought; a disappointment but not a surprise.

Like other pine woods reptiles and amphibians that have been eliminated from large portions of their range, the Southern Hognose Snake has become somewhat of a phantom presence in southern Mississippi. Locals still remember it, and conservationists still search likely areas and go on wild goose hunts based on rumors or phone calls from uncertain sources. Reports of Black Pine Snakes and Dusky Gopher Frogs still emanate from Louisiana's Florida Parishes, and I've met people in Mobile, Alabama, who swear they've seen Eastern Indigo Snakes on weekend outings, but invariably these tips can't be confirmed and only serve to raise false hopes. Likewise, the Southern Hognose Snake remains an absentee from the list of Mississippi and Alabama herpetofauna. I survived my near fatal encounter with fire ants, but the hognose has been vanquished by them. And throughout the Gulf Coast, fire ants continue their domination.

Description

The Southern Hognose Snake is an almost comically short and stubby serpent. The sharply upcurled snout is a hallmark of this species. In its southern congeneric, the Eastern Hognose Snake (*Heterodon platirhinos*), this character is less developed (Conant and Collins 1991). Base color is sandy gray, with a bold pattern of 3 rows of black blotches. One row of blotches (20–28) runs down the center of the dorsum, where the ground color may have a dirty orange hue, while the other two rows of smaller blotches run alongside, with the smaller blotches oblique to the central row. The lower sides are marked with additional smaller and more ragged dark markings. The underside is unmarked white or faintly clouded with gray. Hatchlings are patterned as adults, but in lighter hues, and if any orange wash is present it is feeble. The trend of neonatal snakes having proportionally larger heads than adults is particularly evident in this species (plates 6.8, 6.9).

Heterodon simus is the smallest of the hognose snakes, not exceeding 2 feet (61 cm) (Conant and Collins 1991). It exhibits strong sexual dimorphism, with females growing noticeably longer and more heavy bodied than males. Adult females average 45.5–51 cm, while the more slender males are mature when they attain approximately 35.5 cm in length.

Distribution Notes

According to Tuberville et al. (2000), the current distribution is an assemblage of localities scattered across the Carolinas, Georgia, and northern and central Florida. Based on museum records, currently occupied sites are

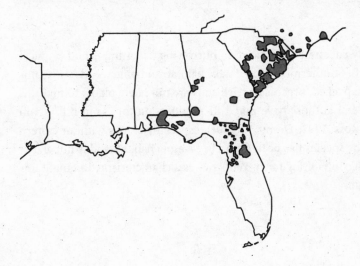

relics of a once more widespread distribution. If Tuberville's estimation is accurate, the decimation of this species across the South is truly shocking. Strongholds still remain, primarily in Richmond County, North Carolina, Marlboro County, South Carolina, and isolated pockets in central Florida.

No recent records have been reported for Alabama or Mississippi. Were it not for the hypothesis that fire ants are a major factor in the species' decline, its absence in these two states would be puzzling because it should continue to co-occur with persistent Black Pine Snake populations there. Tuberville and colleagues proposed no specific mechanism explaining why *H. simus* would be particularly vulnerable to fire ant attack. In light of the extinction of *H. simus* in Mississippi and Alabama on sites where *Pituophis* is still found, it may be that not all snake species are equally threatened by fire ants. One might presume that larger species would be less vulnerable owing to their capacity to eventually exceed the size that fire ants are able to overcome. If this line of reasoning is sound, other small snake species could be predicted to be missing at locations where Southern Hognose Snakes have vanished. There are examples of areas where this is not the case. I find Pine Woods Snakes to be fairly abundant in Mississippi's De Soto National Forest. However, *Rhadinaea* is more secretive and less diurnal than *Heterodon*, and perhaps its different mode of foraging and activity patterns keep it out of harm's way with regard to fire ants. In any case, the hypothesized vulnerability of Southern Hognose Snakes to fire ants may be a consequence of a combination of their morphology and behavior. There may also exist some other ominous and active environmental perversion that we do not yet recognize.

Habitat

Loose, sandy substrate, conducive to burrowing, is an important physical component of Southern Hognose Snake habitat, and pine woods with saturated or well packed, organically rich soils are not likely places to find this species. Several authors have noted its predilection for "sand hill" scrub (Rossi and Rossi 1992; Desmond 2001), especially slopes with an eastern exposure. The species has been found in savanna habitat, and is not an exclusively ridge inhabiting form. Wiregrass is a characteristic dominant understory plant.

Ecology and Natural History

Courtship and mating occur in late spring, but no other detailed observations on wild snakes have been made. A rare and valuable field note was contributed by Neill (1951a), who described two natural mating events that took place in May. Price and Carr (1943) published the only description of a clutch from a wild breeding. Their female laid 6 eggs after an unspecified period of captivity, over a 3-day period in October. Based on the protracted oviposition and the description of the texture and color of the shells, this was likely an infertile clutch, so the October date may not be a typical time for free-ranging *H. simus* to lay. The majority of our knowledge about the reproductive biology of this snake comes from captives. Clutch size ranges from 6 to 14 (Edgren 1955), but 8 eggs comprise a typical clutch (Rossi and Rossi 1992). It is not unusual for captive females to produce two or even three fertile clutches over the course of the summer from a single mating (S. Bowman, pers. comm.). Incubation period is similar to that of most North American colubrids, about 55–60 days. Hatchlings are quite large considering their adult size, averaging 17.8 cm long (Rossi and Rossi 1992; Jensen 1996).

Like *H. platirhinos*, this species is a specialist predator of anurans, especially toads (*Bufo* spp.) and Eastern Spadefoots (*Scaphiopus holbrookii*) (Goin 1947). The Oak Toad likely figures heavily in the diet of *H. simus*. While anurans are important prey items, lizards are also taken. Beane et al. (1998) found remains of Six-lined Racerunners, *Apidoscelis sexlineata*, and a Ground Skink, *Scincella lateralis*, in the stomachs of recently hatched *H. simus*. One adult specimen had eaten a Southern Fence Lizard, *Sceloporus undulatus*. All of Beane et al.'s observations were made in North Carolina.

Although occasionally found in flatwoods and savanna, *H. simus* focuses its activities on the drier ridges. Fortuitous encounters are most often described as forest clearings, or open sandy areas with little or no canopy. This snake is diurnal (Tuberville et al. 2000), exhibiting a bimodal diel activity pattern, with most encounters taking place during early morning or late afternoon (Rossi and Rossi 1992). Snake hunters report that the best way to find Southern Hognose Snakes is by driving up and down low-traffic roads and backwoods sand tracks to encounter roving individuals (Griswold 2000). Owing to their proclivity to exploit forest openings, they may be especially sensitive to the maintenance regime that periodic fire provides.

The species has an extended active season, partly as a result of its tolerance for cooler temperatures. It has been observed basking while temperatures were only slightly above freezing (Desmond 2001). Here again, the role of open ridge environments may be important to the ecology of this species, as the cold weather baskers were situated in microhabitat exposed to full sun made possible by the thin canopy cover afforded by this situation, and typical of it. Deciduous woodland, or pine stands that have been fire suppressed or in need of thinning would provide few such full-sun basking opportunities.

The resemblance of *H. simus* to the Pigmy Rattlesnake is striking. The two species are sympatric throughout the entire range of *H. simus*, and *Sistrurus* often occurs in pinelands. It is tempting to confer mimicry as the underlying mechanism for this similarity, although the idea lacks definitive testing. Being venomous but unlikely deadly to larger predators, the rattlesnake represents an excellent theoretical model for a harmless mimic.

Conservation

Although many species of pine woods herpetofauna are believed to have contracted into smaller areas as once vast distributions have been increasingly fragmented, *H. simus* is one of a few taxa (others being *Lampropeltis getula meansi*, *Lithobates sevosus*, and *Pituophis ruthveni*) where this trend has been documented by historical versus modern collecting data. The splintering of pine woods endemics into disjunct populations is the central concept behind the distribution maps presented in this work. These figures depict a once widespread ecological community now relegated to a limited number of parks, preserves, and miscellaneous categories of lands where, due to a variety of conscious or accidental scenarios, the disruption of pine woods ecology has not been complete.

Heterodon simus is somewhat of a mystery with regard to its abundance. Despite its disappearance from large expanses of formerly occupied territories, the snake is still locally quite abundant in scattered corners of its historic range. Being a diurnal prowler of barren sandhills, the Southern Hognose Snake is a conspicuous species to an observant field biologist, further accentuating the perception that it is common in specific locales. On Internet discussion forums dedicated to snake hunting, several herpetologists regularly report finding multiple specimens during brief one- or two-day hunting excursions. Anyone familiar with snake hunting knows that finding four or five specimens of even a common species of snake in a single day is an event that is unusual and difficult to repeat. No doubt the individuals who boast such impressive feats with *H. simus* are highly skilled field workers who have accumulated a great deal of experience searching for this species, and have become experts on the habits and behaviors of *H. simus*. Much like Ross Allen and his storied ability to divine Short-tailed Snakes without peer, these collectors have developed a sixth sense of when and where to find *H. simus* after years of patient observation. Nonetheless, the ability of these individuals to find so many hognose snakes repeatedly at some locations indicates that the species is still locally common. Equally certain, though, is that throughout most of its sizable historic distribution, the Southern Hognose Snake has vanished. The conundrum remains, and it begs a quick answer before time has run out. Why has the species disappeared in some regions yet remains common in others?

Summary

Startling little is known about this snake, either as a captive or in the wild. It was described to science in 1766, and today's authors continue to struggle when characterizing it. Modern chroniclers often revert to extrapolations from data on the better-known *H. platirhinos*, or paraphrase previous authors' comments that were themselves passed down from paper to paper and originally came from old, stale literature, or present broad generalities in place of specific particulars. The most detailed general ecological information was compiled by Edgren (1955), yet this literature review reveals as many gaps in our understanding as it answers questions.

Short-tailed Snake
Stilosoma extenuatum

The legendary herpetologist Ross Allen was a man of diverse talents. Foremost, he was a showman. He had the gift of holding an audience and educating them under the guise that they were merely being entertained. Other herpetologists, most notably Raymond Ditmars, had earlier pioneered the precedent of sharing herpetological knowledge outside the confines of a cloistered club of academics and professionals. Ross Allen, however, was the first successful herpetologist entertainer, preceding Marlin Perkin's famous television career and the recent flood of herpetologist television personalities that have come into vogue. Those who knew Allen said he was a man.

of great charisma. Ross Allen was a celebrity herpetologist. Allen was also a scientist. That Allen earned scientific respect without a degree is a shining testament to his strong character and inherent gifts. Allen's professional reputation was also significantly bolstered by his long partnership with biologist Wilfred T. Neill, with whom many landmark papers were written. The body of knowledge of the herpetofauna of the Florida pine woods owes much to the lifetime work of Ross Allen and Wilfred T. Neill.

Another of the distinct and rare talents demonstrated by this man was a knack for finding reptiles and amphibians in their habitats, an ability that virtually every herpetologist calls upon regularly but only a handful of people possess to the degree of Allen. To be a good field herpetologist requires not only keen senses, but also an empathy and understanding of the animals of interest. Those so gifted have a certain feel for the creatures they seek. One of the benefits of Allen's skill in finding reptiles was his contribution to our understanding of the Short-tailed Snake.

Although Allen and other exceptional field collectors may appear to have nearly magical abilities to find rare reptiles and amphibians, they share certain qualities that account for much of their seeming good fortune. The pine woods herpetofauna is dominated by species that are notoriously difficult to find. These are animals that many biologists, even those working in the Deep South where these species occur, know only from accounts in guidebooks rather than from firsthand experience in the natural habitat. The Short-tailed Snake is the epitome of such a creature. A breakdown of the characteristics of remarkable field herpetologists is instructive for anyone wishing to increase their chances of finding rare and secretive animals like the Short-tailed Snake.

As often repeated in these species accounts, few pine woods habitats are in a condition that will support the full range of potential herpetofauna. Recognizing quality sites is the prerequisite of any successful search for a habitat-sensitive species. In the case of *Stilosoma*, basic characteristics to look for are open vistas at eye level and conspicuously charred tree trunks, indicating the regular occurrence of fire. Pine forests within the range of *Stilosoma* that appear overgrown with brush will likely be lacking such particular species.

A quality hunting ground must also be large enough to support an ecosystem. Many attractive parcels of longleaf ridge are so small that they are ecologically dysfunctional, and when one spends time surveying these plots they reveal themselves to be like dioramas in a natural history museum, where pristine beauty is only surface deep. Longleaf pine trees, wiregrass, and sandy slopes can persist where the fauna disappeared long ago for lack of suitable ranging area and insufficient resources at the roots of the trophic web. Since the area required by species varies greatly, it's not possible to state an acreage below which no pine woods species can persist. In general, amphibians require less space to maintain viable populations than many reptiles, but some small snakes, such as *Tantilla* and *Diadophis*, can also make do with modest-size preserves. More typically, snakes require large areas of good habitat to maintain their presence. As a very rough guide, from personal experience looking for reptiles and amphibians in areas ranging from residential backyards to vast wilderness in national forests, I have found that isolated sites of fewer than approximately 324 ha are unlikely to contain much of the herpetofauna that would be found in a larger area in similar condition.

Once good-quality habitat has been found, a successful search depends on knowledge of the species' microhabitat preferences. Many pointless hours of searching can be avoided by focusing on the hot spots in a generally favorable area. Places of only a few square meters in area, where multiple components such as moisture, light levels, deadwood, leaf litter, slope, vegetation, abundance of preferred prey, to name just a few considerations, all overlap at optimal levels for one particular species, are where serious searches should be conducted. Finding such sites is a talent in which exceptional snake hunters are exceptionally skilled. Allen focused his search for Short-tailed Snakes in and around Ocala National Forest, a preserve that contains many fine sites for this snake, as well as a multitude of generally similar areas yet lacking in a few key qualities.

Dry, sandy, and sunny were the specific parameters for which Allen looked. Such conditions coexist along the crests of ridges. A ground cover of leaves from xerophyllic oaks is also typical of situations where *Stilosoma* has been collected. However, Ross Allen may have had other signs in his search image that only he knew. He may not have been fully aware of the elements that drew him in for a thorough search, because if a skilled hunter is quizzed about why a successful hunt was launched at one location instead of another, he or she will often respond that there was just something about this particular spot that made it seem that it would yield the quarry. While such statements might be taken as an indicator of a "sixth sense," it is possible that the hunter has subconsciously learned to respond to a subtle but crucial environmental characteristic. If one observes a skilled field worker on the hunt for a specific animal, the pattern of activity will be intensive examination of certain spots, time spent primarily hunched over, studying every detail and object on the ground. These exhaustive searches are interspersed with much longer spells of extended hiking, usually in a random and changing direction, until some feature up ahead or to one side once again seizes the herpetologist's attention and draws him or her in to a rapid beeline, whereupon another meticulous search ensues. The biologist is looking for the favored microhabitat of the species, both consciously and unconsciously. In this way, the field biologist greatly increases the odds of finding the species desired.

Detailed knowledge of a species' natural history and behaviors is key to finding secretive forms. However, with such a poorly understood species as *Stilosoma*, familiarity can be gained only through vast personal experience with the animal and its natural habitat, precious insight that Ross Allen possessed. For Allen, each encounter with *Stilosoma* was a treasure trove of insight into the snake's habits and activities, and with every experience he became more and more attuned to this little known creature. Some of this wisdom he shared, and some he took away with him, a tragic fact since he lived and worked in a long-lost era when central Florida was a true natural wonderland of vast tracts of habitats that are now evaporating away. Some of his firsthand experience in the dry pinelands of Florida could not be relived today.

Description

The Short-tailed Snake's form represents an extreme of morphological adaptation. Compared alongside the adaptations in snout morphology of the pine snakes, the rostral scute specialization of the hognose snake, and the minute dimensions of the crowned snakes, the body form of *Stilosoma* is still quite remarkable. Yet its bizarre form is paradoxical because its systematic relationship leads immediately to the kingsnakes, *Lampropeltis*, from which it seems not far removed. Despite its physical divergence from familiar snake patterns of body form, *Stilosoma* is merely a highly specialized kingsnake derivative, relatively recently evolved from its parent stock.

The immediately striking feature of the Short-tailed Snake is its greatly elongated and remarkably slender body. Adults average 35.5–51 cm in total length (Conant and Collins 1991), and are not known to exceed 66 cm. Even large adults are no greater in diameter than a pencil, and a distinct lack of anterior and posterior taper accentuates the stringy appearance. Despite its common name, the Short-tailed Snake is more accurately described as a "long-bodied" snake, because its tail is normally proportioned for a typical snake with similar head dimensions. However, because of the extended length of the snake, the tail accounts for only 9–12 percent of the length of the body (Brown 1890).

The pattern of both adults and juveniles consists of 50–80 irregularly shaped dorsal blotches that may be ovoid or squarish in shape and wider than long. A second row of smaller blotches is located along each side. The blotches are dark chocolate brown in color. Ground color is variable, ranging from silvery gray to pale tan, often with a faint orange wash. On the middorsal scales, the orange color, when present, is concentrated such that a rusty streak, interrupted by the dorsal blotches, is created. The venter is light gray heavily marked with black blotches that extend onto the lowest dorsal scales to create a sublateral row of markings (plates 6.10, 6.11).

The head bears a distinctive marking, it being a Y-shaped dark brown design with the forks pointing anteriorly. This marking is very reminiscent of both putative sister taxa (Dowling and Maxson 1990), the Milk Snake, *Lampropeltis triangulum*, and even more so, the Mole Kingsnake, *L. calligaster* (Stejneger 1894).

Head scalation and morphology (Dowling and Maxson 1990) exhibit characteristics indicative of the species' fossorial nature. Internasal scales are often fused with the prefrontal scales. The loreal scale, and usually the preoculars, are absent. The 2 parietal scutes are unusually large and long. The overall effect of these missing scutes and fusions is a reduced lateral head scalation and increased contact between the dorsal head scutes and the supralabial scales atop the line of the mouth. The head is markedly depressed, and this may be a character linked to the reduced scutellation on the sides of the head. The snout is also distinctly flattened and extends well past the terminus of the lower jaw, an obvious advantage when pushing through loose sandy substrate.

The taxonomic position of such a divergent serpent form has been studied and debated since the genus was described (Brown 1890). Its superficial resemblance to kingsnakes was doubted to be informative by Auffenberg (1963), whose paleological study of Florida snakes included an examination of the vertebrae morphology of extinct and recent *Stilosoma*. Auffenberg reported on several characteristics of the vertebrae of Short-tailed Snakes, citing their overall flattened shape, short broad condyle, lack of distinct separation of the paradiapophyses, epizygopophyses without spines, and abbreviated neural spine as sharply divergent from other Nearctic colubrids and collectively indicative of only a distant relationship with the genus *Lampropeltis*.

In contrast, Dowling and Maxson (1990) used immunological data to support an argument that superficial resemblance between *Stilosoma* and the kingsnakes is in fact the result of a close relationship and relatively recent divergence. Using albumin amino acid sequences, these authors interpreted a difference of 1:7 amino acids between any two species as indicating approximately one million years of independent evolution. *Stilosoma* exhibited an albumin amino acid difference with *Lampropeltis* samples of 6 amino acids. Based on these data and method of interpretation, Short-tailed Snakes and kingsnakes were derived from a common ancestor around the Pliocene, during a period of high sea level, when central Florida was broken into an accumulation of islands. This hypothesis is not weakened by the confirmation that fossilized snake remains clearly referable to the genus *Stilosoma* date from the paleo-Florida islands during the mid-Pliocene (Auffenberg 1963).

Highton (1956) described geographic variation in this taxon and defined three subspecies. Specimens exhibiting the greatest extent of head scale fusion occur in the eastern portions of the range. On these snakes, prefrontal and internasal scutes are consistently fused, and more often than not so are 2 or more anterior infralabials. This eastern population exhibits relatively low body blotch counts of 68 or fewer. *Stilosoma* from elsewhere in the distribution do not exhibit the distinctive head scute fusions, but Highton was able to perceive a subdivision within this sample on the basis of body blotch count, one group with 68 or fewer blotches, and another grouping with 69 or more blotches. Highton proposed that three mensural and morphological subsets be designated as subspecies. The eastern population exhibiting low blotch count and head scute fusions were named *S. e. extenuatum*, the population to the north exhibiting typical head scutellation and high body blotch counts was designated *S. e. multistictum*, and snakes hailing from the southern portion of the range, with lower body blotch counts (distinguishing them from *S. e. multistictum*) and separate internasals and infralabials (distinguishing them from the nominate subspecies) were labeled *S. e. arenicola*.

With a slow trickle of specimens over the ensuing years following Highton's descriptions, it appears that the taxonomic distinction of the populations not exhibiting head scute fusions is unwarranted. At the time of

Highton's study, no *Stilosoma* from Hillsborough County were available for examination. Woolfenden (1962) reported on two specimens collected from Hillsborough County that prompted him to reconsider the diagnostic characters purported to define three subspecies. Both Hillsborough specimens lacked scute fusions, thus did not meet the criteria for *S. e. extenuatum*. Based on collecting locality, these snakes were predicted to have fewer than 68 body blotches and be referable to *S. e. arenicola*. However, both specimens exhibited high blotch counts (77 and 82.5), thus violating Highton's three subspecies concept. Head scutellation characteristics did fit Highton's hypothesis, thus confirming a geographically consistent difference on that basis. Woolfenden (1962) concluded that subdividing populations possessing unfused scutes into two subspecies based on body blotch count should be discontinued, but supported a modification of Highton's view of *Stilosoma* as consisting of two distinct entities dividing the total distribution east and west.

Distribution Notes

The Short-tailed Snake is a Florida endemic. Its distribution overlies two distinct ecoregions, the xeric pine scrub of north central Florida, and the Florida Ridges. Locality records extend from central Columbia County south to southern Highlands County, and from Orange and Seminole counties east to the central Gulf coast (Campbell 1978).

As with several other taxa discussed that have similar ranges around the Florida Ridges, *Stilosoma* is a primitive form. One hypothesis holds that the ancestral stock evolved in isolation on the islands that gave genesis to the

physiographic features seen today in the form of great stretches of north-to-south-running high sandy ridges. With the subsidence of high sea levels during the Pleistocene, the distribution of *Stilosoma* expanded and the morphological and meristic variation discernible today evolved in populations isolated by more recent physical features. Nonetheless, the Short-tailed Snake's many distinctive features are probably reflective of the primitive characters from which it and its sister taxon, kingsnakes of the genus *Lampropeltis*, are derived.

Habitat

The habitat is a crisply dry land where dead oak leaves crunch underfoot and scrubby brush and a sparse contingent of pines do little to shield the ground from direct sun. A unifying feature of sites supporting the snake is the nature of the substrate, which is composed of one of several highly leached, nutrient-poor sands of a fine enough grade to be conducive to burrowing by such a small creature.

The diverse physiographic provinces of Florida result in the three named subspecies occupying slightly different habitats. The northern and western taxa, *multistictum* and *arenicola*, occur in areas of typical Florida forest and scrub, with canopies of pine of varying densities and a conspicuous underlying component of xerophyllic oaks. Forest litter provides abundant surface cover in these habitats. The nominate subspecies also occurs in such habitats, but its range includes the central Florida ridges, thus it is found in the most xeric environmental extremes tolerated by the species.

The spatial stratum occupied by *Stilosoma* is the topmost layer of leaf litter, pine straw, and other debris. Because it is an uncommon species, no technique for finding it has more than slim odds of being successful, but in theory, meticulously sifting through fallen oak leaves along open ridges would yield the best results. Warm rains that temporarily quench and cool the parched landscape may stimulate brief surface activity, and *Stilosoma*, particularly juveniles, are sometimes found prowling out in the open during warm, sunny mornings after a night of heavy rain.

As with many of the smaller and most secretive members of the pine woods herpetofauna, the microhabitat of *S. extenuatum* is more complex than it appears to the casual eye. Campbell (1978) stated that the taxon might have distinct preferences for specific soil compositions. He related that in trials using captives, a preference was observed for three varieties of naturally occurring substrate material, all of them fine, loose, and with a high sand content.

Ecology and Natural History

Considering the morphological traits that suggest an interesting natural history, it is disappointing that an ecological summary of *Stilosoma* must be so brief. The practical difficulties of utilizing techniques such as radiotelemetry to investigate its natural history are so daunting that it is not reckless to predict that such will never be completed. The species' apparent scarcity and likely fossorial tendency renders standard drift fence and pitfall trap arrays poorly suited for population sampling of *Stilosoma*, except perhaps during brief periods in the season when surface activity and long-distance moves might occur (Campbell 1978). Coupling the impracticality of research with the realization that the central Florida pine ridges are steadily disappearing leads to the depressing conclusion that we may never fully understand the interactions of this snake with its environment.

The best-known aspect of *Stilosoma* natural history is its dietary preferences and feeding behavior. Short-tailed Snake biology has been clearly glimpsed only for captive specimens, and feeding behavior and reproduction are two areas where an observant herpetoculturist can make important contributions to our knowledge of rare and secretive reptiles. Even early authors quickly realized that this small, burrowing species, so closely resembling kingsnakes, likewise preyed on snakes. They noted that lizards and newborn mice, items that typically would be devoured by kingsnakes of similar dimensions, were flatly refused by captive *Stilosoma* (Ditmars 1939; Allen and Neill 1953). Thus, here is an example of an important insight into the fundamental nature of a snake virtually impossible to observe in the wild that was gained through the perceptive eyes of reptile keepers. The first of these reports came from the great Florida naturalist Archie Carr. In a brief note, Carr (1934) encapsulated the nature of *Stilosoma* dietary ecology so completely that little has been added in the 70 years that have followed. Carr's contribution to our knowledge of this genus is that it avidly feeds on Florida Crowned Snakes, *Tantilla relicta*, and thus the Short-tailed Snake's feeding habits are succinctly characterized. Mushinsky (1984) built on Carr's note by capitalizing on his rare good fortune of possessing a captive *Stilosoma* and subjecting it to feeding trials consisting of five lizard and four snake species that are sympatric. Mushinsky's specimen refused to feed on any of these potential prey except for *T. relicta*. There are several facultative ophiophages among the snake fauna of the United States, including *Lampropeltis*, *Drymarchon*, *Masticophis*, *Micrurus*, *Micruroides*, and *Agkistrodon piscivorous*, but no other ophiophagous snake in North America is known to be single species specific.

Stilosoma dispatches *Tantilla* by constriction, although it is a protracted process. Both Carr and Mushinsky described feeding bouts that required more than two hours for Short-tailed Snakes to kill their prey. On occasion, especially if the prey is small, *Stilosoma* will forgo constriction and swallow a crowned snake that is still alive and struggling (Rossi and Rossi 1993).

An interesting series of observations by herpetoculturists Rossi and Rossi (1993) suggest that *Stilosoma* is a more general ophiophage prior to reaching adulthood. These authors offered a variety of snake species to a yearling specimen and reported that it consumed all five species: the Brown Snake, *Storeria dekayi*, Redbelly Snake, *S. occipitomaculata*, Lined Snake, *Tropidoclonion lineatum*, Rough Earth Snake, *Virginia striatula*, and Ringneck Snake, *Diadophis punctatus*. The acceptance by *Stilosoma* of *Tropidoclonion* is surprising, as that taxon is not sympatric with *Stilosoma*. The Rossis pointed out that their successful feeding trials involved a variety of readily available species, and included both living and prefrozen, thawed dead offerings. They believed this indicated that *Stilosoma* would be well suited for captive maintenance by a dedicated keeper. Indeed, if we are to ever learn about other aspects of the Short-tailed Snake's biology, such as breeding behavior, clutch size, and hatchling appearance and morphology, it will most likely be from specimens maintained in labs, zoos, or private collections.

The degree to which the Short-tailed Snake relies on a single species as its prey is very unusual. Few native snakes specialize on a single prey item. Narrowing this trophic ecology further is the fact that *Stilosoma*'s prey is itself a highly specialized predator. The predator-prey linkage in this chain, from beetle larva to *Stilosoma*, is a remarkable example of extreme ecological specialization. It is a reminder of how special and different the high dry ridge habitat is, and of its dramatic evolutionary consequences for the organisms that survive there.

Conservation

There is no thoughtful opposing viewpoint that contends this species is not precariously pre-adapted to be extremely vulnerable to habitat disturbances. The upland pine habitats of central Florida are not as extensive as they were even as recently as the mid-twentieth century when Ross Allen roamed these sites, yet sizable tracts do remain where *Stilosoma* can still be found. The question lacking a clear answer is whether the Short-tailed Snake is stable and secure in its present-day haunts. A decisive answer to this urgent question, lacking any long-term population studies or comparable short-term surveys separated by a span of decades, is currently beyond the grasp of conservation biologists.

Summary

The consequence of exceptionally narrow specialization is an increased vulnerability to environmental change. The more fixed specializations a species possesses, the greater the number of environmental factors that, if altered, will exert stresses on resident individuals. For a species such as Slowinski's Corn Snake, which feeds on a variety of endotherms and likely some amphibians and invertebrates, an environmental perturbation would be unlikely to sever all connections with the prey base, and local corn snakes might need only to shift diet composition to exploit the most abundant component and survive a challenge. Corn snakes are physically well constructed to scale trees or live on the forest floor, and occupy both situations routinely, further broadening their ability to persist in habitat that has undergone major structural change. A species like Slowinski's Corn Snake has few "Achilles' heels" that would lead to its local downfall. *Stilosoma* is situated at the opposite position in terms of being able to adjust to environmental changes. It is a species that has evolved to function extremely well within a precise set of environmental and ecological parameters. Change a single component and the ripples of consequence will be felt by the population and threaten its survival. The physical and behavioral traits of *Stilosoma* are the observable indications of a commitment to a rigid ecological position, and result in a species that exists in a fragile state.

Eastern Pine Snakes
Pituophis melanoleucus lodingi and *P. m. mugitus*

Captive breeding contributes to the conservation of reptile and amphibian species, but it can also be a source of damage. Both attributes are illustrated by the history of the two southern pine woods races of *Pituophis melanoleucus* in captivity. The herpetological staff of zoos were the pioneers of the new field of herpetoculture during the heady early years of the 1960s and early '70s. In those days, the husbandry and breeding of captive reptiles as a studied art was the exclusive provenance of zoological parks and a handful of privately operating individuals—most of whom were associated with zoos in some fashion. The expansion of this small group of dedicated technicians into the huge hobby and business it has become began in the mid-1970s. With zoos now shifting focus to in situ initiatives, most of the advances and significant achievements now take place in privately owned collections. Rather than a small group of zoo curators and zookeepers being the sole source of captive-bred reptiles and limiting the distribution of offspring to members of the club, there are now hundreds of very talented and productive hobbyists producing vast numbers of specimens in their homes and disseminating them into the international marketplace. Rather than hundreds of hobbyists making pilgrimages to herpetoculture conferences to listen to a revered zoo curator expound on his latest breeding success, it is now the zoo curators who follow the lead of the private sector, studying their accomplishments with envy and amazement, as that community continually pushes the limits of possibilities with regard to captive reptile and amphibian husbandry.

Prior to the birth of modern herpetoculture, many reptile species commonly kept in captivity today were unobtainable. Some of today's most popular and available species, including Gray-banded Kingsnakes (*Lampropeltis alterna*), Ringed Pythons (*Liasis boa*), Dumeril's Boas (*Acrantophis dumerili*), and Prehensile-tailed Skinks (*Corucia zebrata*), were, in recent memory, almost unheard of as captives, and when they did become available, they were so expensive that only well-financed zoos could afford them. That all of the above species are now frequently occupying the collections of beginners is a testament to the effectiveness of advancements made in the field of herpetoculture over the past several decades.

Two more examples of the transformation from a rarely seen taxon of legendary rarity to a bread-and-butter staple of the reptile-breeding hobby are the Southern Pine Snake, *P. m. mugitus*, and the Black Pine Snake, *P. m. lodingi*. Particularly with respect to the Black Pine Snake, the progress made through the annals of captive culture—from obscurity to boom and then bust—is a chronicle of the positives and negatives of reptile conservation via herpetoculture.

The Black Pine Snake reproduced for the first time in captivity at the Philadelphia Zoo (Kevin Bowler, pers. comm.). Since that time, many zoos have reproduced the species, to a fault, one could say. The first private herpetologist to reproduce the taxon was Terry Vandeventer, based in Jackson, Mississippi. A small number of these early hatchlings found their way into other private breeders' hands, ushering in a period when, for the first time in the 50-plus years since the snake's discovery, a hobbyist or zoo curator could aspire to acquire specimens with a realistic chance of success.

Prompted by the availability of a small number of captive-bred hatchlings, a growing interest and demand ensued. The entrepreneurial element of herpetoculture recognized a good opportunity with this rare and beautiful snake, one for which there was a high demand and a potentially steady but limited supply (as a result of the subspecies' ease of reproduction but low fecundity). Demand soon outstripped the supply, and prices increased. With money to be made, some collectors turned their efforts toward obtaining wild-caught stock, a bit of a long shot given how few specimens had been documented since the subspecies was described, but now worth the effort. Two methods were pursued. At least one local pet store in Mobile, Alabama, engaged in a brisk business by serving as middleman between local residents who occasionally encountered Black Pine Snakes on rural roadways, and enterprising collectors eager to get a headstart advantage on their competitors by acquiring snakes ready to breed and produce marketable young immediately. The shop cultivated contacts with the field collectors and the herpetocultural community and sold as many snakes as they could acquire at impressive mark-ups. The second, more direct avenue was to place "wanted" posters at key gathering points, such as rural convenience stores, timber company offices, and hunting club camps, offering a cash reward for uninjured snakes. Some people even got creative, printing reward notices on antiqued parchment with old-fashioned lettering so as to resemble the "wanted" posters of the Old West, and offered bounties of $100–$150 per specimen. Through these two strategies, a sizable captive founder stock was established, resulting in more breedings and increased availability.

It is impossible to know how many specimens were collected during the 10 or so years these practices were in force, but judging from the frequency of adults appearing on dealers' lists and the growing number of hobbyist breedings, the number may have exceeded 100.

The economic rules of supply and demand are good predictors of general interest in a particular reptile species by the pet or hobbyist trade. While this phenomenon is especially applicable to the community of hobbyists and commercial suppliers who support their acquisition of specimens and cost of maintenance through the sale of surplus, zoos, or more precisely, the staff who manage the herpetocultural collections of zoos, are not immune to the influence of rarity, value, and perceived desirability.

After several years of heavy production, with established breeders attempting to get clutches from every female and bringing young specimens to maturity as quickly as possible, and while new caretakers snapped up every available hatchling at premium prices or anxiously got on waiting lists for the opportunity to do so, the desirability of the Black Pine Snake began to fade. As more and more collectors maintained and bred the snake, a glut of surplus hatchlings was presented in the marketplace each fall, sales grew sluggish, and prices fell. Before long, facilities both private and public began dispersing their adults and ceased working with the subspecies, citing an inability to place hatchlings and an unwillingness to keep a snake when there seemed no good purpose in it. The number of zoos holding specimens is a reflection of this trajectory and decline, standing at 22 in 1987 (Slavens 1988), prior to the breeding boom, to a high of 51 in 1991 (Reichling 1992), but falling again to 22 by 2005 (International Species Inventory Program 2005). Unsteady commitment to the captive-breeding effort prompted the elimination of the formal studbook for the race (Reichling 2000), because there was nothing to manage.

While zoos were demonstrating their waning interest in pine snakes by eliminating them from their inventories, some in the private sector embraced a new use for the taxa. The new utility contrasted unfavorably with the earlier motivation of maintaining and reproducing a threatened and beautiful snake from the Mississippi and Alabama pinelands. With a growing interest in so-called "designer" reptiles—the highly altered products of selective breeding to achieve colors and patterns not found in nature—some enterprising genetic engineers reasoned that the gene for melanism in *P. m. lodingi* and the propensity of *P. m. mugitus* to produce occasional variants with reduced or absent patterns offered the potential to develop immaculate white *Pituophis*. A pure white pine snake, oddly and inexplicably, is a highly desirable product, though the precedence for coveting such a creation can be seen in the strange affinity that zoos and the general public alike share for mutant white tigers and alligators. For a time, the desirability of captive-bred Black and Southern Pine Snakes was revived, but many of these specimens were destined to be paired with their conspecific subspecies counterpart rather than with their own form.

The new market was satiated with the initial acquisitions of pure *lodingi* and *mugitus*, whereupon continued generations of pure subspecies became irrelevant to the goal. Today, there exist a wide variety of undeniably beautiful color phases and pattern aberrations in a variety of native snake species, including pine snakes, and fewer offerings of naturally occurring forms. This would pose no risk for these subspecies in the wild were it not for the fact that the breeders and dealers sometimes lose track of pedigrees. At the Memphis Zoo, I have been offered snakes that were sincerely believed to be pure *P. m. lodingi* and *P. m. mugitus* by their owners, but it was clear that they were not. This has happened repeatedly and attests to how many hybrid crosses have circulated unknowingly among collectors and breeders. Coupled with this problem is the practice of releasing unwanted pets or, in some cases, the intentional introduction into what is perceived to be the ancestral habitat. Again, on multiple occasions, I have been called upon to identify *Pituophis* collected in Tennessee, Mississippi, and Louisiana only to determine that the snake in question was a product of captive herpetoculture and not genetically benign to the native congenerics it might have encountered and reproduced with. By this practice, the captive breeding of pine snakes has come full circle, from the ethically unsound consumption of delicate wild populations, to conservation of these forms by maintaining self-sustaining populations and eliminating the need for wild harvest, to once again threatening natural forms by producing undocumented and uncontrolled mongrel snakes. Most disturbingly, the pursuit of the beauty displayed by snakes through selective breeding for color and pattern lessens appreciation for the spectacular creations of nature by relegating them to the ordinary and plain, which they are not.

Description

Adult Black Pine Snakes are solid black from head to tail. The only consistent marking anywhere on the body is a pure white spot, or several spots or blotches, on the gular region, crisply set off from the black throat as though a small dab of white enamel paint had been carefully placed there. On some individuals, the labials display a hint of dark brown hue that may also extend further onto the head. The venter is deepest black, usually unmarked but occasionally with extensive white down the center, especially on specimens from the eastern end of the distribution. The ventral scales are polished glass-smooth and display iridescent purple and blue highlights on freshly molted snakes. The subcaudal region is usually a slightly lighter color than the jet-black ventrals. Freshly shed adults are stunning creatures owing to their strongly keeled middorsal scales that catch the light and cause the snake to glisten when it moves (plate 6.12).

Hatchlings are patterned, but to a variable degree. The darkest neonates display faint traces of pattern only along the posterior flanks and tail, where faint outlines of blotches can be seen. Occasional hatchlings exhibit a weakly defined pattern from head to tail, such that accurate blotch counts can be made. Body blotch counts for *P. m. lodingi* are in the range of 25–27.

A fairly rapid ontogenetic melanism occurs in Black Pine Snakes, and upon attaining a length of 80 cm most are patternless. Subadults less than 80 cm vary in the extent of pattern they retain, depending on their age, the number of ecdyses that have occurred, and probably their genetic predisposition for this character.

Adult *P. m. mugitus* closely resemble the nominate subspecies, with which they intergrade along the Peidmont in Alabama, Georgia, and South Carolina (Conant and Collins 1991). They reliably differ from *P. m. melanoleucus* only in the details of coloration. Southern Pine Snakes average 25 dorsal blotches (Reichling 1995), as do northern pines, but in *P. m. mugitus* the markings are ill defined owing to the muted coloration and poor contrast between pattern and ground color. Ground color is variable but usually a somber earthen shade, ranging from dirty gray to light muddy brown. Occasional snakes have very light or nearly white ground color as is seen regularly in *P. m. melanoleucus*, but even in these lighter animals, the blotches are not set off starkly as they are on the northern subspecies. Rarely, Southern Pines have been encountered that lack most of their pattern and are nearly solid white or pinkish, and these are highly sought by hobbyists for selective captive breeding, as were a handful of pure white, leucistic mutants that surfaced some years ago (plate 6.13).

Hatchling Southern Pines are patterned and colored similarly to adults, except that the subtlety of hues—tans, yellows, variations of brown, etc.—do not develop until several sheds have taken place, and the pattern contrasts more sharply with the ground color.

Hatchlings of both subspecies are comparable in size, and quite large. Although snake neonate dimensions are correlated to egg size, which in turn is negatively correlated to clutch size, so data for a particular taxon will be a generalization, most neonate pine snakes range in total length from 45 to 55 cm (Reichling 1990).

Distribution Notes

Based on an assembly of historical records, *P. m. lodingi* once ranged across a swath of Gulf coastal territory from Louisiana's Florida Parishes to coastal southwestern Alabama and all of southern Mississippi, wherever high dry pineland held sway (Conant 1956). This region was originally dominated by longleaf ridge and savanna habitat. The natural geographic barrier on the western front of the distribution was the Mississippi River and associated bottomlands, separating the Black Pine Snake from the Louisiana Pine Snake, *Pituophis ruthveni*, for the past 12,000 years. To the east, a zone of unfavorable habitat was created by the presence of the Mobile and Alabama rivers, although some gene flow between the two subspecies apparently takes place or occurred until recently, as indicated by pine snakes in Escambia County, Florida, and Alabama east of Mobile Bay that are intermediate in color and pattern between *P. m. lodingi* and *P. m. mugitus*. These intergrades have jet-black markings and much black stippling along the flanks, yet also exhibit white pigment in the interspaces around the blotches that leaves the pattern clearly visible even on large adults.

The current distribution of the Black Pine Snake is greatly constricted from its former expanse. Extant populations are known to remain in 9 of the 14 Mississippi counties where older records exist (Duran 1998b). The once extensive but now marginally intact pine woods in southwestern Mississippi, represented primarily by the holdings in Homochitto National Forest, may no longer support pine snakes, as the taxon has not been documented there since the late 1970s. These snakes, from the western limit of their

range, were not completely melanistic and displayed a discernible blotched pattern dorsally. The strongholds of this snake in Mississippi—areas where suitable habitat exists in extensive, protected stands of pine and where the snakes are uniformly patternless black as adults—are in Harrison, Stone, Perry, and Wayne counties.

In Alabama, extant populations persist in all counties where historical records have been compiled, but this statement is misleading as the areas of occurrence are considerably reduced. Counties where Black Pine Snake phenotypes have been documented are Clarke, Washington, and Mobile. Most of the piney landscape delineated by these county borders is completely unsuitable for endemic herpetofauna, and devoid of pine snakes. Sites where pine snakes persist are few, small, and scattered within this general area.

No pine snakes have been observed in southeastern Louisiana for decades, and the Black Pine Snake, along with the Dusky Gopher Frog, is a pine ridge endemic probably long extirpated from the state.

The distribution of *P. m. mugitus* is known with less precision than that of *P. m. lodingi*. The form occurs sporadically across a large geographic area, from southeastern South Carolina through southeastern Georgia, and the northern three quarters of the Florida peninsula and the entire panhandle. Despite this sizable territory, as with other pine habitat endemics, areas of habitat suitable for it are often few and far between; thus the Southern Pine Snake is one of the best examples of a pine woods species whose general distribution creates an inaccurate image of its actual scarcity. The Southern

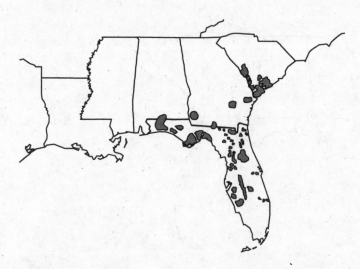

Pine Snake also intergrades with its two regional congenerics over broad areas of range contact.

Habitat

Pockets of optimal pine snake habitat still exist throughout the range of the species, but they are becoming increasingly hard to find. Some of the best and most extensive samples remain in central Florida. The most important of these areas is the spectacular Ocala National Forest. Here the pine woods are so vast that exploring them gives a real sense of what natural Florida must have been like up until 50 or 60 years ago. In these precious tracts, a once commonplace but now rare combination of vegetation, geology, and forest dynamics maintains a beautiful variation of southern pineland.

All of the excellent sites for supporting populations of *P. m. lodingi* or *P. m. mugitus* share three essential characteristics. A sandy or sand substrate is the most commonly occurring necessity for pine snakes, as it is the one most impervious to destruction by human activity. It is the hallmark of any collection site for these two snakes that they be seen atop or beneath such substrate.

While all good pine snake habitats must be located on very sandy soils, there are many sites within the overall distribution of the species that meet this criterion but lack one or both of two other key elements. The most subtle habitat element required by pine snakes is subterranean refugia. Although they are uniquely adapted by head morphology and osteology to burrow directly into the soil (Knight 1986), they do this mainly when preparing chambers in which to lay their eggs. At other times, pine snakes utilize the burrows made by other animals such as Eastern Pocket Gophers and Gopher Tortoises, or pre-existing retreats such as burnt-out stumps and root systems or the crevices created by uprooted trees (Duran 1998a). Contrary to popular belief, there is no such thing as a "snake hole" in the sense that any native species actually constructs tunnels underground in which it lives. Pine snakes occupy burrows and holes going deep underground, but these features must already be present. Such things exist only in finely tuned pine ridge ecosystems, where natural tree falls are left undisturbed and fires regularly sweep through and destroy some trees. These sites have a rich and complete biotic component, including obligate burrowers like gophers and tortoises, and these are places where pine snakes are to be found.

The third critical element for pine snake habitat is open midstory, allowing saturated sun exposure of the forest floor. One explanation for the importance of this character is its pivotal role in the presence of pocket gophers, which are a primary prey item of the southern subspecies. Without ample sun at the ground level, grasses will not thrive, and these are the main food base for pocket gophers. Although *P. m. lodingi* has not been historically sympatric with pocket gophers west of Mobile Bay, most Mississippi specimens for which detailed collecting data are available have been found prowling by day in open, sunny scrub. In more mesic and lush areas, or where fire has been suppressed, Black Pine Snakes are rarely found. Since these populations are known to feed on a variety of rodents not dependent on fire-driven forest structure, the snakes are frequenting such habitat for other, as yet unknown reasons (plate 6.14).

Ecology and Natural History

Until the advent of radiotelemetry equipment small enough to be utilized on snakes, the activities of species such as the pine snake were invisible to biologists. The only time specimens were encountered and observed was when the paths of roaming snake and human intersected by rare happenstance. A pine snake slowly trailing its way across some remote jeep track on a spring morning in the southern pine woods is a beautiful sight, but hardly informative. Only after herpetologists began implanting radio devices into wild specimens so they could be tracked and examined on any day desired were they able to peer into the mostly hidden lives of these secretive animals.

In most respects, the ecology and natural histories of the two southeastern races of pine snake do not greatly differ, and a general synopsis applicable to both can be presented. Both forms are secretive and partially fossorial. On the rare occasions when time is spent on the surface, it is always during daylight hours. Despite the infrequency of sightings, pine snakes make forays over the surface of considerable distance during certain times of the year. Home ranges are large, typically 50 hectares or greater, with males traversing roughly twice the area as females over the course of a year (Duran 1998b). As a result of this combination of diurnal activity and impressive mobility, one of the most frequent settings for human encounters with pine snakes is on rural paved roads traversing good habitat. Peak surface activity is during the spring, but a second pulse of movement occurs in the fall. The spring activity peak occurs during April and May, and most specimens found during this period are males roaming in search of females. The late season activity peak commences in late summer, beginning in August, when females are more frequently encountered, much as with the pattern seen in the Louisiana Pine Snake.

During the intervals separating these periods of activity, pine snakes spend their time in underground refugia. Pine snakes are the largest North American snakes that are fossorial. Duran (1998a), who followed 13 radio-tagged *P. m. lodingi* in the Camp Shelby military training base in Mississippi, found that his specimens were underground 66 percent of the time. This is considerably less time spent in underground refugia than studies have revealed for the Louisiana Pine Snake, which appears to have an even more narrowly focused natural history.

Both subspecies are sympatric with the Gopher Tortoise. Many organisms, ranging from invertebrates to mammals, are highly dependent on the deep burrows that tortoises excavate, and have consequently developed commensal relationships with this species. Surprisingly, pine snakes, particularly the Black Pine Snake, do not appear to make frequent use of these features. Even where Gopher Tortoises are abundant, Black Pine Snakes rarely enter the burrows, preferring instead to shelter in crevices and crannies associated with the root systems of rotting pine stumps (Duran 1998a). These are abundant structures in natural ridge habitat owing to hurricane and other wind-related damage, as well as disease and erosion causing tree falls. Here then is another example of the contrast between the devastation that tropical storms can wreak on human lives and the beneficial purpose they serve in pineland community ecology.

In most respects, distinguishing the natural histories of the two pine wood endemic subspecies of *P. melanoleucus* is futile. Dramatic phenotypical diagnostics do not extend into the ecological interactions displayed by these two forms, with but one clear exception. Pine snakes, including the distinctive Louisiana sister species, have long been known to prey almost exclusively on fossorial rodents, especially pocket gophers (Conant 1956). Gophers are abundant co-inhabitants in many xeric pine woods associations. However, neither species of pocket gopher native to the southeastern United States occurs on the portion of coastal plain between the Mississippi and Mobile rivers (Pembleton and Williams 1978; Sulentich et al. 1991), creating a gap in distribution that overlaps the majority of the distribution of the Black Pine Snake. Only east of Mobile Bay, in Baldwin County, Alabama, do records of *P. m. lodingi* and pocket gophers coincide. Thus it would appear that the menu for *P. m. lodingi* differs from that of *P. m. mugitus* and *P. ruthveni* (Vandeventer and Young 1989). Indeed, when Duran conducted his study of *P. m. lodingi* at Camp Shelby, he found the prey taken by this snake encompassed a variety of small mammals, the majority being Cotton Rats, *Sigmodon hispidus*, (Duran 1998a). Duran speculated that the small population of Black Pine Snakes in Baldwin County, where pocket gophers co-occur, was more centered around the rodents and their burrows than for consubspecifics elsewhere in the region.

The Southern Pine Snake is sympatric with the pocket gopher throughout its distribution. It feeds heavily on these mammals and is commonly found sheltered in the rodents' tunnel systems (Franz 1992); like the Louisiana Pine Snake, it may prefer these structures as hibernation refuges (Neill 1948). The Southern Pine Snake's natural history mirrors what is known of the Louisiana species with respect to its interactions with its environment, even though the two taxa are situated at opposite ends of the longleaf pine belt.

Conservation

The Black Pine Snake is threatened by habitat destruction (Cliburn 1980; Reichling 1986). Jennings and Fritts (1983) cited forest management practices, expanding urbanization, and the indiscriminate killing of these conspicuous diurnal snakes as factors in their apparent decline, and they considered the subspecies uncommon throughout most of its range, although locally more abundant at some locations. Foresters interviewed during Jennings' and Fritts' status survey agreed that *P. m. lodingi* was less common than it once was. The snake is now protected in all states where it occurs.

That the Black Pine Snake still occurs in Mississippi is almost entirely credited to the protection and management its habitat enjoys in De Soto National Forest. Almost three quarters of the confirmed sightings and collections made since the late 1970s have been within this tract of public land (Duran 2000). Another public preserve, the Marion County Wildlife Management Area, also supports the taxon. The balance of recent records comprises primarily single specimens collected in scattered private holdings, some too small to support self-sustaining populations.

In Alabama, there is no extensive acreage set aside from development as there is in Mississippi. Although the three counties from which Black Pine Snakes were historically known still support the subspecies, the actual areas of occurrence have shrunk dramatically. The sites where snakes are currently found are small and disjunct privately owned parcels, and their long-term status is quite uncertain. These conditions predict a gradual decline in the number of locations where Black Pine Snakes will persist in Alabama. *Pituophis m. mugitus*, with its more extensive geographic range, is more secure than *P. m. lodingi*. It is, however, listed as a Species of Special Concern by the State of Florida.

Summary

Recognition that the Eastern Pine Snakes are in peril has waned as captive breeding successes have made these taxa familiar offerings in the commercial pet market. A consequence of this is the fading interest among zoos to dedicate space to these snakes. Perceived rarity drives demand in the reptile hobby, and the drop in prices for captive-bred pine snakes since the 1980s is in part a function of the faulty presumption that if the species is abundant in herpetoculture, it cannot be as rare in the wild as once thought.

In truth, these two snakes are rare and continually under pressure in a shrinking habitat (Vandeventer and Young 1989). Both are no longer found in many areas that once regularly produced specimens. Pine woods to the west and southwest of Mobile, Alabama, near towns such as Irvington and Grand Bay, where the first examples of the subspecies were collected and where it was commonly seen for decades after, are quite simply and profoundly gone, vanished, never to return. Given the Black Pine Snake's proclivity to shelter within root channels and holes created by decaying pine stumps, and its preference for open, sunny habitat with sparse canopy cover, one can picture a suitable landscape created by low-intensity commercial silviculture operations. Clearcuts would be kept small and scattered, and balanced by an abundance of acreage supporting pine forests maintained thin and open through selective harvests. Site preparation for new crop plantings would exclude roller chopping and stump removal. Under such a management regime, the Black Pine Snake could probably coexist with pine agriculture. This scenario was the norm in many pine plantations in decades past, and burning was vigorously applied as well. Under the much more intensive procedures common today, where stands are kept dense and cut on short rotations in order to maximize profitability per acre, and where acres once cleared are further insulted by churning of the substrate and removal of all stumps and other refugia, the prognosis for the Black Pine Snake will continue to be grave.

Louisiana Pine Snake
Pituophis ruthveni

A biologist would immediately know that something special was happening upon entering Louisiana Pine Snake habitat. In appearance it is reminiscent of other pine ridge situations, but its setting is unexpected. Driving over the country roads that sparsely criss-cross Bienville, Natchitoches, and Vernon parishes reveals mostly routine landscape. Most of the acreage in this region supports nondescript woods supporting a generic mix of oaks, hickories, sweet gum, and some pines, and the intermittent bayous that are a regional cliché. Passing through for the first time in 1982, I wondered where these snakes could ever have been found.

The type description and subsequent early papers mentioned these parishes and the little towns—Bellwood, Cypress, Lucky, and Hodge—over and over, but I saw nothing to immediately suggest that any pine ridge endemic could survive here. I was only a few kilometers from Lucky, once a significant way station for the early loggers but now merely a crossroad of two rural highways, where several snakes had recently been collected. Although a shift in dominance from deciduous brush to pines indicated I was entering a degraded pine ridge community, the brown organic soil and dense underbrush bore no suggestion of the open, sandyland scrub that pine snakes prefer. Then, suddenly, as the road inclined slightly up a slope, the entire scene changed dramatically. Underbrush had dissipated and open patches of grayish white sand were prevalent. Driving more slowly, I saw prickly pear cactus growing in low patches. A Roadrunner darted across the road, further accentuating the feeling of being in some strange desert habitat where the vegetation wasn't quite right. Loblolly plantations, so common elsewhere in Louisiana, were excluded here because of the infertile dry soil; this was still longleaf country.

As I rounded a curve and dropped a few meters in elevation, the scene quickly returned to a more generic, mixed vegetation, and I was back in the world I'd left only a few minutes before. This was my first and most memorable encounter with the land of the Louisiana Pine Snake. Its discrete home is precariously situated, surrounded by an expanse of inhospitable land pushing against its boundaries and swallowing it up wherever human activity upsets the delicate biological interactions.

Description

Adult Louisiana Pine Snakes, their scales dull and rough like pine cones, rich but somberly hued in black, brown, and russet, and carrying the scars of dozens of subterranean battles with prey, are a clear reflection of life for a snake in the southern pine woods. The species generally presents as a tan snake marked with 28–38 darker blotches that become better defined posteriorly. However, these are highly variable snakes, a fact not recognized for many years after their discovery, probably because of the limited opportunity available to herpetologists to examine a sufficient sample. Features considered diagnostic include bold ventral marking, postocular stripe, and brownish hue, but these show enough variation and overlap with congenerics to render them unreliable as distinguishers. Body blotch count is the best superficial physical characteristic to consistently distinguish this species from congenerics. Eastern Pine Snakes (*P. melanoleucus*) have fewer body blotches (23–30), while Bullsnakes and Gopher Snakes (*P. catenifer*) always have higher counts (40–79) than *P. ruthveni*. Mean body blotch count for *P. ruthveni* is 34, ranging from 28 to 38 (Reichling 1995) (plates 6.15, 6.16).

Adults always display a striking disparity between anterior and posterior pattern. The light interspaces separating body blotches on the forward third are so heavily infused with dark speckling that the margins of the blotches can be completely obscured, rendering them nearly uncountable. Toward midbody this character subsides rather abruptly, so that over the rear third the blotches are cleanly set off from the light ground color. If the front and back portions of a specimen are lined up alongside each other, they appear to be from completely different kinds of snakes.

Ground color is generally a shade of light to medium brown, quite distinctive from the other two pine woods congenerics. Variations in hue do occur, however, the most common being a tendency toward straw yellow. Frequently some reddish tones encroach into the posterior pattern and, rarely, the entire snake is awash in rich red or orange. I have seen three snakes with black blotching on an almost white ground color, and these were very reminiscent of Northern Pine Snakes, *P. m. melanoleucus*. Each of these pigments matches a dominant substrate color in various parts of the snake's range, but no correlation of snakes matching the local conditions is apparent. More likely, potentially adaptive polychromatism appears to be present throughout the various populations.

The venter is typically covered with extensive black markings of irregular shape that congeal to produce a heavy black suffusion. A smaller proportion of specimens possess only a light (usually reddish) maculation in the ventral scales, and occasional snakes have unmarked scutes except for small spots along the margins.

Distribution Notes

Seven disjunct tracts of relict longleaf ridge habitat in central Louisiana and extreme east Texas represent the present distribution of this snake. The most robust population lives on corporately owned land utilized as industrial forest in Bienville Parish. A small portion of this population's range enters the Winn District of Kisatchie National Forest in Natchitoches Parish. A second population occurs in the Kisatchie Hills region of Kisatchie National Forest in south central Natchitoches Parish, including the vicinity of Red Dirt Wildlife Management Area. The third Louisiana population, which appears to be disconnected from the previously described site (Rudolph et al. 2006), occurs primarily within Peason Ridge military training grounds, and probably into small portions of adjacent Kisatchie National Forest. The fourth population in Louisiana is centered on Fort Polk military reservation, with possible incursions into adjacent Kisatchie National Forest along its southern border. There is a slim possibility of a few additional, extremely small, populations elsewhere. Two such areas are Beauregard and Calcasieu parishes.

In Texas, three very tenuous populations are known in Angelina National Forest and commercial timberland located nearby, and each site is appar-

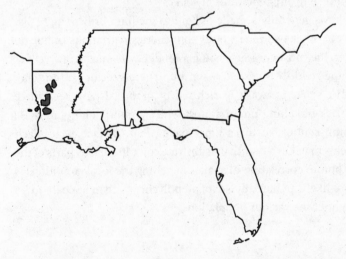

ently separated from the others by unsuitable habitat. A single specimen from Wood County (Williams and Cordes 1996) documents a recently extirpated population.

Smith and Kennedy (1951) considered Bullsnakes (*P. catenifer sayi*) and Louisiana Pine Snakes to be parapatric, and contended that a zone of intergradation existed in east Texas. Despite their confident assertion that contact between the two forms "leaves relatively little to be assumed and [the] direction [of their] conclusions incontrovertible," there is actually no evidence of contact between these two taxa in recent times. Two specimens, from Caldwell County, were suggested to be indicative of a zone of intergradation. Wilks (1962) identified one of these snakes as a Bullsnake and the other as a Louisiana Pine Snake. Thomas et al. (1976) discussed the putative pine snake specimen and the controversy it created, and determined that it, also, was a Bullsnake. Although it is impossible to prove that no *Pituophis* occur in the area of contention, considering how fragmented the current distribution of even marginally suitable habitat tracts is in east Texas, it is more certain now than ever before that Louisiana Pine Snakes are completely isolated from congenerics.

Habitat

Ideal habitat consists of xeric, sandy-soiled pine ridges historically dominated by fire-climax longleaf pine-oak vegetative associations and supporting abundant pocket gophers. On infrequent occasions, the species has been found in pine-oak situations, especially at the northern extremes of its distribution. This snake avoids pine plantations, especially tracts with extensive clear-cuts. This habitat association may be more correlative than direct, because pocket gophers—the prey species upon which adult *P. ruthveni* specialize—prefer the same situations. The grassy understory characteristic of fire-climax longleaf ridges is essential to pocket gophers because they feed almost exclusively on the roots and young shoots of this vegetation. Longleaf communities subjected to periodic fires support thriving pocket gopher populations and, in turn, healthy populations of Louisiana Pine Snakes. So, although no reptile is more representative of the pine woods in Louisiana and east Texas, the snake is probably oblivious to the presence or absence of any particular species of tree. The Louisiana Pine Snake is tied to this type of forest because sandy soil and sun-baked grasslands are ideal habitat for pocket gophers; longleaf pines are simply the best-adapted tree for surviving in these dry, fire-swept areas, and thus are closely correlated with the presence of pine snakes.

It is difficult to know whether the Louisiana Pine Snake is a scrub or forest species because its habitat has been dramatically altered over the past century. Many scrubby areas where snakes occur are the result of recent reclamation and restoration efforts, and are neither historic nor stable. Driven by the mandate to expand suitable habitat for the endangered Red-cockaded Woodpecker, federal and state conservation agents are remodeling some loblolly forests long denied fire back into a semblance of their original condition. This effort has taken place only over the last decade or so and, as a result, these sites support young trees and open canopies, imparting a scrubby character. Historical accounts from prelogging days depict a vast swath of ancient longleaf pines, with great columnar trunks supporting a nearly closed canopy high overhead, shading the forest floor (Boyer 1979; Frost 1993; Early 2004). A hint of such forests can be felt at only a few spots these days, such as in sections 5 and 6 of the Winn District of Kisatchie National Forest, Winn Parish, Louisiana. Many of the places Louisiana Pine Snakes once inhabited may have looked like this, and thus the species' occurrence in sporadic pioneer scrubs may not reflect its true nature.

Ecology and Natural History

The reproductive behaviors of *Pituophis ruthveni* in nature have remained inaccessible to observation. Only one gravid female has been collected from the wild, a 132 cm specimen that subsequently laid 5 eggs on July 3. Using the period between copulation and oviposition recorded for captive specimens as a reference, and the date of oviposition for the wild-caught snake, wild mating likely takes place during the latter half of May. This is also the period when most captive copulations are observed. A series of females encountered during the 1980s in August were noted to be very thin, suggesting they had recently oviposited (Young and Vandeventer 1988). Many of the apparently "spent" females were traversing watermelon patches (which are common locally) suggesting that these open, sandy sites are favored nesting areas. Burger and Zappalorti (1991) observed similar nesting site choices by Northern Pine Snakes in the New Jersey Pine Barrens. However, if these puny specimens really are females fresh from nesting, it indicates that this species sometimes lays later in the year than other snakes in the region.

Most reproductive data come from observations made on captives. The species exhibits several remarkable distinctions. No other snake in North

America lays eggs as large (4 x 13 cm), nor does any native colubrid (except tiny fossorial taxa such as the Florida Crowned Snake and Pine Woods Snake) produce such small clutches. Twenty-two clutches had a mean number of 4.8 eggs (mode = 4). The range for clutch size from this sample was 3–9, because occasional captive-reared specimens, growing much larger than any specimen recorded from the wild, are capable of laying unusually large clutches. As would be expected from such large eggs, hatchlings are enormous (55 cm or longer) and on average larger than any other North American snake (Reichling 1988, 1990). The large hatchlings may be in response to pressure for rapid growth to a size where pocket gophers can be eaten, or the large eggs may be selected for because of their greater resistance to desiccation in the quickly draining sandy substrate, and large hatchling size is merely a nonadaptive consequence (plate 6.17).

This snake's ecology is inextricably tied to pocket gophers. This statement could not have been made prior to 1995, when the first reports from D. Craig Rudolph, Shirley Burgdorf (U.S. Forest Service, Southern Research Station, Nacogdoches, Texas), and colleagues began appearing. Representing what is unquestionably the most important insight into this species made to date, Rudolph, Burgdorf, and team established through a series of interesting studies that the Baird's Pocket Gopher, *Geomys breviceps*, and its burrows are the key to understanding habitat choice, diet composition, feeding strategy, activity patterns, home range, and perhaps even aspects of the reproductive biology of this snake. Also significant is John Himes' radiotelemetry study of 18 snakes, which contributed an understanding of the movements, behavior, and habitat requirements. The ecological overview that follows is drawn heavily from the pioneering work of these individuals.

Louisiana Pine Snakes spend the majority of time below ground. Above ground activity is often in the immediate vicinity of pocket gopher burrows, into which the snakes flee when disturbed. A typical encounter with a Louisiana Pine Snake consists of walking into a sunny patch of sandy ground and catching a momentary glimpse of a snake in motion as it retreats down a nearby hole. Experience teaches that, when searching for pine snakes, the first thing to look for is an area where pocket gopher burrows are concentrated. Baird's Pocket Gophers cluster in some spots, yet are scarce or absent in surrounding areas that appear to offer similar habitat. Studies of movements of pine snakes implanted with radio transmitters have disclosed that they seldom stray far from these gopher hot spots. Annual home range may be large, up to 300 ha for males (Himes 1998), but most activity is concentrated in much smaller core areas where gopher densities are high. The overall wide-ranging areas utilized by these snakes is a function of their need to make short-duration, long-distance moves to sites supporting numerous gophers. Himes (1998) reported that pine snake activity, both seasonally and daily, is bimodal. He noted that daily activity peaked in late morning and again in late afternoon, with no activity observed between 2000 and 0500 h (plate 6.18).

A feeding behavior used by some fossorial snakes is particularly characteristic of this species. The technique consists of using the side of the body to press prey against a firm surface, such as a burrow wall, until it suffocates. The Rudolph-Burgdorf team demonstrated that this behavior is not characteristic of all *Pituophis* spp. (Rudolph et al. 2002). In trials comparing prey-handling strategies of Louisiana Pine Snakes, Bullsnakes, Black Pine Snakes, and Western Rat Snakes, only Louisiana Pine Snakes and Gopher Snakes exhibited press constriction when presented with pocket gophers. Black Pine Snakes and rat snakes either killed gophers by conventional constriction or refused to subdue pocket gophers at all. The authors noted that this feeding strategy is advantageous to snakes, which prey chiefly on large, burrow-dwelling rodents. Although press constriction is effective in dispatching rodents in the tight confines of an underground tunnel, it may not be as efficient as typical constriction in which a snake completely encircles its prey with coils. With standard constriction, a rodent's ability to twist and bite is thwarted by the enveloping coils. Observations on captive Louisiana Pine Snakes reveal that rodents are able to gnaw on their attacker for considerable time before suffocation; that this also occurs in the wild is borne out by the observation that most adult Louisiana Pine Snakes carry numerous scars.

As is true for its congeners, encountering a Louisiana Pine Snake on the surface is an unforgettable experience, accentuated in this case by the unequaled rarity of such an event. The defensive shams of an alarmed snake are awe inspiring. The foreparts are drawn into a series of S-shaped coils and elevated high off the ground in readiness to strike. At the same time, explosive, phlegmy hisses are blasted at the threat with each exhalation. Snakes of this genus all share this unique ability due to their unique adaptation of a vertical blade of epiglottal cartilage, which amplifies and adds a menacing quality to the sound. The mouth is held slightly open during these displays, giving further warning that the snake intends to inflict a bite if the offending danger comes within reach. In common with many snakes, the Louisiana Pine Snake brings the tail into play, shaking it and producing a faint rattling sound if it comes in contact with debris. Even the facial countenance of these snakes adds to the effectiveness of the defense, at least to human eyes, because the heavy supraocular scales hooding the slightly recessed eyes give the snake an enraged expression, even when in repose. Although a pine snake in the midst of such a display will strike and bite energetically, it will usually quiet down immediately if scooped up and held loosely. At this time, specimens often display another characteristic trait of being extremely squirmy (Conant and Collins 1991) and constantly trying to wriggle out of grip.

Conservation

The Louisiana Pine Snake is undoubtedly one of the most endangered vertebrates in the United States (Rudolph et al. 2006). But because the snake spends the majority of its time below the surface, another possibility that must be considered is that the snake is not rare at all, just rarely seen. In response, it should be pointed out that only 178 specimens have been documented since its description more than 70 years ago. Furthermore, many areas where snakes are known to have occurred are now completely altered and unsuitable for them. Areas that do support these snakes are limited to seven widely separated patches of pine ridge, several of which may be too small to support a self-sustaining population.

After synthesizing their conclusions that this species is associated with habitat that supports pocket gophers, Rudolph and Burgdorf (1997) advanced a hypothesis to explain its recent disappearance throughout much of its former range. With a significant portion of the land within the snake's historic distribution now converted to commercial pine plantation, the summer wildfires that were once integral to shaping the ecosystem are prevented, and undergrowth is controlled with infrequent winter backfires or by the application of herbicides. Neither of these methods prevents the eventual buildup of a brushy midstory. When the midstory becomes dense enough, it suppresses and eventually eliminates the grassy understory that is the main food source for pocket gophers, leading to their decline. It appears, Rudolph and Burgdorf contend, that without fires, pocket gopher populations dwindle, and without pocket gophers, there can be few if any Louisiana Pine Snakes.

Summary

Large and beautifully patterned, a spectacular sight when aroused in defense, and possessing a unique assemblage of adaptations to a peculiar and threatened habitat, the Louisiana Pine Snake may well become the first U.S. snake species to exist solely in zoos. Over the past several decades, the species has contracted into a handful of isolated sites—tiny fragments of Texas and Louisiana pine woods whose widely scattered locations testify to a habitat that was once far more extensive. Some of these parcels are so small, the equivalent of a few city blocks, or so marginal in quality, that it's likely only a few old stragglers are keeping these spots on the list of active sites. Just try to find good pine ridge habitat in Louisiana; it's nearly as elusive as the Black Pine Snakes, Gopher Tortoises, and Dusky Gopher Frogs that are following their maternal woodlands into extinction by just a couple of steps.

Florida Crowned Snakes
Tantilla relicta relicta and *T. r. neilli*

One of the arguments advanced by those who oppose the preservation of large tracts of pine woods in their natural state is that the threatened species contained therein could be adequately conserved through captive breeding, and that the species could be perpetuated indefinitely in zoos, where people could better appreciate them. Why should human usage of these lands be hampered by concern over a salamander or snake, they argue, when the population could be removed and placed in zoos? Zoos themselves, although they represent the camp that believes in protecting natural areas and restoring disturbed habitat to pristine condition, have contributed ammunition to this proposal by touting their utility as refuges for species in decline, so that biodiversity can persist in captivity if not in the wild. Zoos strive to attain this goal through managed programs like Species Survival Plans (SSPs) and pedigree studbooks, which are designed to husband small founding populations without genetic bottlenecks or inbreeding depression for centuries to come. The concept is appealing, and has been promoted persistently through the metaphoric description of zoos as modern-day Noah's Arks, the Great Flood of the present day represented by habitat destruction. Perhaps one day, the healthy populations of endangered species maintained by zoos can be set down safely on the friendly shores of protected reserves that were not in existence during the "great flood." Although the practice of maintaining assurance colonies of endangered species has a role to play in conservation biology, it is not an antidote for the problems faced by many species, and some animals are simply unsuited for large-scale captive cultivation. Furthermore, unlike its biblical namesake, the zoo ark has only so much room available and cannot carry every species to safety even if the skill were there to enable such a grand plan.

Because of the constraints exerted by limited space, money, and personnel, zoos must choose which species will receive the bulk of these resources, leaving many threatened animals without managed programs. Zoo biologists take great pains to develop and follow rational decision-making processes that identify the best candidates for intensive long-term management. Despite the quest for objectivity, human biases are inevitably at play, as evidenced by the skewing of programs for vertebrates rather than invertebrates, and with most vertebrate programs dealing with mammals, followed by birds, then reptiles, and finally amphibians. In 2004, the American Association of Zoos and Aquariums listed 69 SSPs for mammals, 21 for birds, 9 for reptiles, 2 for amphibians, and single programs each for fish and invertebrates (American Association of Zoos and Aquariums 2005). Judged by the proportion of Earth's biodiversity given to these classes, or by the number of imperiled species each contains, there is no logical explanation for this bias for mammals; human aesthetics have led to this outcome. The same is true of private hobbyists and breeders, who tend to choose species based on their potential to generate demand and profitability, and all but ignore small, drab, wormlike snakes such as *Rhadinaea*, *Diadophis*, and *Tantilla* species. As a result of these prejudices, threatened species such as the Florida Crowned Snake have little hope of receiving much attention from herpetoculturists of any stripe.

Even if the Florida Crowned Snake were targeted as a conservation priority by zoos, there would be many serious obstacles to block success. Although husbandry techniques for snakes have come a long way over the past four decades or so, Florida Crowned Snakes are among the many species that are still considered extremely difficult to maintain. Their unusual dietary preferences present a daunting challenge to anyone wanting to keep more than a handful of captives, because the prey are themselves without any standardized care protocols and so would have to be laboriously collected from the wild as needed. Tiny fossorial snakes such as *Tantilla* have narrow tolerance limits for ambient moisture, rendering them highly prone to both desiccation or skin infections should the substrate stray too far toward either end of the soil moisture spectrum.

The fossorial nature of crowned snakes makes the requisite close scrutiny of captives problematic. Delicate snakes such as this need to be examined frequently to evaluate their health and correct problems before serious conditions arise, but the substrate that would be essential for providing a secure medium in which to burrow would also hide them from their caretaker's view. Specimens would have to be dug up regularly, which would stress the snakes and disrupt the security they seek.

To maintain genetic diversity in a long-term captive population, the typical lifespan of the species is an influential factor. Because some alleles are lost to the progeny at each reproduction, genetic heterozygosity in a closed captive population is in part a function of the number of generational turnovers within a given span of time. Thus, given the same number of founders, populations of species that can be maintained for decades before they need to produce progeny replacements will conserve initial genetic diversity better than a population of a species with a short generation time. Crowned snakes, because of the artifact of delicacy in captivity and perhaps also because of an inherently short lifespan, present the captive management disadvantage of a short generation span. Anecdotal information on a variety of small, fossorial colubrids leads to the inference that the natural lifespan of *Tantilla* is less than 10 years, much shorter than most snake species when well cared for in captivity. With individuals living no more than a decade at best, a small captive population would require new reproductive recruitment at a more rapid rate to maintain stability, compared with a larger, hardier snake species that lived for a more typical 20–25 years in captivity.

Tantilla's low fecundity further exacerbates the likelihood of a poor outcome of a captive rescue or assurance colony. Snakes with higher fecundity than *Tantilla* can replace population mortalities more efficiently with fewer breedings. Quicker turnover rate of individuals in a small population of Florida Crowned Snakes would require frequent breedings to maintain level population size over the long haul, increasing the cost of the program in terms of personnel time and husbandry expertise as well as equipment and rearing cost.

With these considerations in mind, the unfortunate conclusion is that the Florida Crowned Snake will never be a species maintained in captivity in sufficient numbers to insulate it from extinction should the wild populations decline. As support of this contention, one has only to consider the closely related Rim Rock Crowned Snake, *T. oolitica*, a critically imperiled species and similarly ill suited for captivity. Despite the urgent need for action and the theoretical value of a captive breeding program, no such effort exists nor has it been seriously considered. For species such as *T. relicta*, the best hope for its surviving the decades of intensive land development that lie ahead for the central Florida ridges is the sequestering and rigorous natural habitat maintenance of sufficiently large tracts of land, a scenario that seems unlikely at the present time.

Description

This is a tiny, slender snake, overall light tan in color with black markings on the head and nape. In both pine woods subspecies, the dorsum of the body is pale tan and the venter an immaculate white, but they differ in details of their black head patterns. The nominate subspecies exhibits a black collar on the nape created by a break in the black coloration on the first two scale rows. This light neck ring is usually disrupted at the midline. *Tantilla r. neilli* lacks the light markings with the black area or if present, only in the form of a light spot on the posterior portion of the parietal scales (plate 6.19).

Telford (1966) noted differences in head morphology between *T. r. relicta* and *T. r. neilli*, stating that *relicta* has a pointy snout and countersunk jaw, presumably adaptations for burrowing, while *neilli* lacks these modifications.

Distribution Notes

Tantilla r. relicta is largely confined to the central Florida ridge country, beginning in Marion County and terminating at the south in Highlands County (Conant and Collins 1991). These areas represent the oldest territory continually above sea level in present-day Florida. *Tantilla r. relicta* probably occupied a continuous narrow corridor of xeric pine ridge scrub before drastic environmental alterations forever changed this land. Today, the distribution of extant populations resembles a dotted line, where some public and mostly private holdings have managed to preserve vestiges of this environment. Two somewhat anomalously positioned locales are also known to support Florida Crowned Snakes, these being disjunct sites in Collier, Charlotte, and Sarasota counties in southwestern Florida.

Tantilla r. neilli occurs to the north of the bulk of the area occupied by *T. r. relicta*, in dry, sandy situations such as those found abundantly in Ocala National Forest. The taxon has been taken in northern Florida very close to

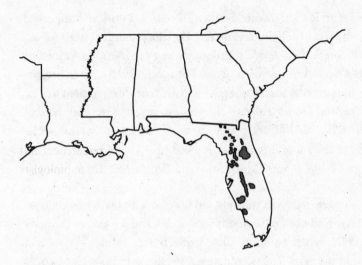

the border with Georgia and, like the Florida Worm Lizard, probably occurs in that state, most likely in the driest uplands of Echols County. That portion of the state of Georgia still promises interesting discoveries to anyone willing to conduct thorough sampling efforts.

Habitat

Tantilla r. relicta and *T. r. neilli* differ somewhat in habitat preferences. *Tantilla r. relicta* typically favors the open, sunny conditions afforded by scrub habitat, particularly the Florida endemic Rosemary Scrub association on the Lake Wales Ridge. In contrast, *T. r. neilli* often inhabits situations where the canopy is more dense and the understory more complex.

Telford (1966) proposed that natural selection has favored head patterns in *T. relicta* according to habitat characteristics and fixed the traits that now help diagnose the two pine woods subspecies. By observing captives, Telford discovered that *T. relicta* basks while lying just below the surface of the sand, with only its head exposed. Telford thought that on bare sandy substrate, a broken head pattern might be more cryptic than a solid marking by disrupting the outline of the head. However, a large black area as found on the solid-marked *T. r. neilli* would act as a more efficient absorber of infrared rays. Since *T. r. neilli* favors less open situations and is afforded more cover by the grasses and low-lying ground vegetation of ridge forests, the more thermoefficient solid black head pattern does not impart a selective disadvantage as it hypothetically does for *T. r. relicta*. Whether or not Telford's untested hypothesis that the pros and cons of cryptic versus thermoefficient head pigmentation in different microenvironments have driven divergent evolution in *T. relicta* is correct, his discovery of the covered basking behavior of these *Tantilla* is an exceedingly rare glimpse into the nature of these secretive and oft overlooked creatures, and is a perfect example of the intricate natural histories that remain beyond our skills of perception.

Ecology and Natural History

Most of what is known of this species is owed to a pair of researchers and their single published report (Mushinsky and Witz 1993). These biologists conducted a 10-year sampling effort of small herpetofauna in Florida pine scrub. The life history sketch that follows is drawn heavily from their research. Florida Crowned Snakes are active during all months of the year, but peak activity occurs April through October, and is at its lowest ebb in January and February. Males and females are encountered at equal frequency. No observations on courtship or mating appear in the literature, but based on Mushinsky and Witz's findings, it can be inferred to take place from March through May. Nearly three quarters of the large sample of gravid females captured were found from late April through early June. Hatchlings appear in May and June, and measure 88–95 mm SVL.

The dietary composition of *Tantilla relicta* is among the narrowest possible. Some native *Tantilla* species may be fairly nonspecific in the invertebrate prey they consume. However, this appears not to apply to *T. relicta*, which we now know from a thorough examination of a large sample of stomach contents. The overwhelming majority of prey items identified in the stomachs of 124 crowned snakes were tenebrionid beetles, of one species and only the larval stage (Smith 1982). This amazingly specialized ecological niche is but one link in an incredible and incompletely known trophic chain, because *T. relicta* is believed to be the primary, possibly the sole, natural prey of the Short-tailed Snake. What predators and prey specializations may lie further up or down this chain are unknown at present.

This is similar to the situation involving the Louisiana Pine Snake, which as an adult preys almost exclusively on the local species of pocket gopher. The pine snake is now endangered, apparently as a result of forest management changes that secondarily impact the snake by depleting its prey. Both the Florida Crowned Snake and the Short-tailed Snake are extremely vulnerable to this sort of insidious indirect threat.

Mushinsky (1984), contemplating the trophic chain connecting *Stilosoma*, *Tantilla*, and Tenebrionidae, reasoned that such extremely restricted ecological interrelation must indicate exceptionally long-standing environmental stability. Mushinsky's insight emphasizes that although *T. relicta* and *Stilosoma* inhabit pine ridge scrub situations, it is not the floristic component that provides the essential parameters necessary for the species, as this aspect of the landscape is prone to dramatic change over great spans of time. Rather, it is the unique qualities imparted by the physical structure of the forest floor and substratum on the high sandy land that have remained static since the mid-Pliocene and allowed such unique reptile taxa to evolve and ecologically intertwine.

The species of the genus *Tantilla*, represented within the scope of this book by *T. relicta* and *T. oolitica*, exhibit one of the lowest fecundities of any North American snake. *Tantilla* species are one of only a handful of taxa known to lay single eggs as a normal clutch, although 2–3 is more typical of the genus (Mount 1975).

Mushinsky and Witz (1993) found *T. r. relicta* both abundant and uniformly distributed throughout their study site, which ranged from fire-suppressed to annually burned plots. Thus the Florida Crowned Snake stands as the only southern pine woods endemic snake for which fire has been demonstrated to be irrelevant. Perhaps this ability to remain unaffected by variations in fire regime is founded on traits similar to those previously discussed for *Eumeces egregius*. Given the extremely narrow diet composition of *T. relicta*, one could assume that tenebrionid beetles are also unaffected by the extent of fire on the local scale. Key to understanding the ecological parameters conducive to *T. relicta* is a thorough understanding of the same factors during the life history of this single insect family.

Conservation

As the smallest pine woods snake, *T. relicta* possesses an advantage over other endemic species in the struggle to survive in a shrinking environment. Although no direct studies on home range have been published for *T. relicta*, it is intuitive that such a diminutive and secretive creature would utilize a smaller area during the course of its lifetime than a larger species. Tiny species such as crowned snakes are more susceptible to overheating if exposed to direct sun than larger snakes, and may also be more at risk of desiccation. Large or medium-sized snakes feed on vertebrates as adults, prey that is often at relatively low density and frequently on the move within large home ranges itself, so the snakes must be equally as mobile during the course of their season of activity. Conversely, a *Tantilla* could conceivably encounter all of the small invertebrates it required for sustenance under the cover of a few discarded boards or fallen palm fronds. With such limited spatial needs, *Tantilla* populations could persist within preserves of very modest dimensions. A small (40 ha) patch of longleaf sandhill scrub, bounded by highways and residential development, yielded more than 500 specimens of *T. r. relicta*, proving the species to be relatively common in comparison with other xeric pine upland snakes (Mushinsky and Witz 1993). Even small city parks, if properly maintained to sustain a natural flora and fauna, could serve as suitable refuges for *T. relicta*. In one sense then, no other pine woods endemic reptile is better suited to survive the consumption of the majority of its habitat so long as appropriate management of a collection of embedded reserves is instigated. This is fortuitous, since captive culture offers no hope for its conservation.

Summary

Crowned snakes are exceptionally difficult to study, and without an understanding of their behavior and natural history, it is impossible to determine whether they have been negatively impacted by habitat fragmentation and alteration. Conservation biologists know that pine woods amphibians such as the Pine Barrens Treefrog and the Dusky Gopher Frog have become imperiled in large part as a result of the vulnerabilities imparted by their reproductive biology and the peculiar nature of the sites where they choose to breed. In contrast, biologists know nothing of the conditions where Florida Crowned Snakes choose to deposit their eggs, though we can be certain that the conditions required are quite specific and possibly difficult to find in the disheveled remnants of our southern pine forests. Similarly, we can make reasonable assessments of refuges that offer potential safe harbor for introduced populations of Louisiana Pine Snakes because their home range requirements have been revealed through laborious fieldwork, information that is completely lacking for crowned snakes. Such vast deficiencies in understanding thwart any accurate evaluation of the rangewide population status of secretive species like *Tantilla relicta*.

7

Rockland Specialists

Key Ringneck Snake
Diadophis punctatus acricus

The contrast between the delicacy of the Key Ringneck Snake and its harsh habitat is a striking juxtaposition. The taxon's natural habitat is relegated to tiny patches, enclosed on some sides by the sea, and blocked on all remaining flanks by an ever-growing human population and its attendant bustle. The boundaries of the assortment of public and private reserves are like the invisible walls of a fortification—constructed of legislation and owner agendas rather than of brick and mortar. Standing at these transition lines, one can almost hear the cracking and moaning of the wall as the forces just outside strain against it, forces that would overrun the protected areas in weeks were it not for the foresight that established and maintains them. Within the fortress walls, a diminutive and fragile snake survives in an environment more stern and restrictive than any experienced by its near relatives living abundantly throughout the eastern United States. Living under the miserly shade of scattered pine trees, among leathery skinned cacti and spiny palms—all seemingly far better suited to the hot, unbaked landscape—the Key Ringneck Snake seeks the limited places where sufficient cool and moisture enable its survival.

The Key Ringneck Snake's relationship with the humans that have so profoundly reshaped its world is not a simple one. In contrast to the stark conditions offered by the pine rockland and the restrictions these place on ringneck occurrence and activities, these snakes find critical environmental parameters more abundant in the tropical gardens that are everywhere to be found in the backyards of Big Pine Key neighborhoods. Thick layers of mulch kept moist by irrigation, decorative brick walkways with deep fissures extending to the inviting cooler layers of soil, constant shade of transplanted palms and fig trees that keep surface temperatures mild during the hottest days, and a protective blanket of leaf litter scattered by foreign ornamental plantings all combine to produce a gentle microenvironment for litter-dwelling reptiles and their invertebrate prey. The development of south Florida's real estate paradise has damaged the unique pineland habitat of the Key Ringneck Snake, while simultaneously providing safe harbor in an alternative environment. How this clash of influences ultimately affects the species will depend on achieving a balance between land development and natural area preservation, and by not permitting the former to subsume what little remains of the pine rockland scrub.

Description

The Key Ringneck Snake differs dramatically in appearance compared with other ringnecks of the southeast. It is the only eastern form lacking the namesake marking. On some specimens, a hint of a neck ring can be discerned as a faded, smudged remnant, but typically there is no trace of it. Hatchlings have not been described, but the phenomenon of the expression of basal phenotypes in neonatal reptiles and amphibians leads to speculation that recently hatched *D. p. acricus* exhibit a ringed neck that fades with each successive shed. Its southeastern conspecifics, the Southern Ringneck Snake, *D. p. punctatus*, and the Mississippi Ringneck Snake, *D. p. stictogenys*, have discrete black spots on the labials (Conant and Collins 1991) but such markings are faint or absent in *D. p. acricus* (Weaver and Christman 1978). The dorsum can be dirty gray, light brown, or black. The anterior ventrum is lightly pigmented and without strong pattern elements (Paulson 1966). Mainland *D. punctatus* spp. average 25–36 cm in length, and may reach 50 cm on occasion, but no information on the range of length specific to the Florida Key subspecies is available (plate 7.1).

Distribution Notes

This subspecies has the smallest range of any U.S. snake taxon. It is known only from Big Pine Key, and there, only in sections exclusive of estuarine marshland or low areas prone to frequent inundation. Terrain appearing similar to Big Pine Key rockland exists on a few of the nearby smaller keys, particularly No Name and Little Pine keys, but supporting data to indicate established populations of the ringneck are tenuous. In addition, *Diadophis* occasionally turn up in Key West, but these specimens appear to be main-

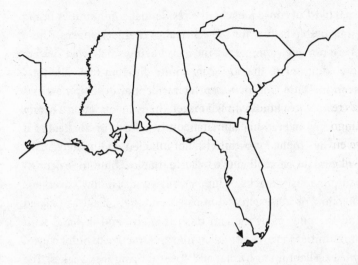

land *D. punctatus punctatus*, and were probably introduced by the vigorous trade of soils, mulches, and garden plants from mainland nurseries.

Habitat

The pine rocklands that *D. p. acricus* inhabits are predominantly an arid habitat interspersed with scattered pockets of mesic refugia. The snake is hypodispersed within this terrain, such that no specimens are to be found over the majority of the acreage encompassed by their range. The snake is best sought in locations where moisture and shade are consistently available, as this diminutive form is not equipped to move great distances over hostile terrain if its microhabitat becomes unsuitable (plate 7.2).

Both natural and artificial environments can provide good conditions for the Key Ringneck Snake. Interspersed throughout undisturbed rockland are low areas, the result of disintegration of the underlying limestone bedrock at or very near the surface. Many of these depressions are not deep enough to develop into solution holes, but they do accumulate leaf litter that settles and decays into a layer of organic soil that is not common in the landscape generally. The organic soil supports a richer flora than is typical in adjacent sites, and the combined effect of these processes are tiny pockets of habitat affording good microenvironments for small, semifossorial snakes. Semitropical hardwood hammocks, the consequence of a different genesis, are not typical in Big Pine Key and neighboring keys, but some sites approaching this condition are present, and these too provide excellent habitat for the Key Ringneck Snake.

In an unusual twist of consequences, the residential communities of Big Pine Key are a favorable feature for the survival of ringneck snakes. These developments are not environmentally intrusive, being small, well spaced, and ecofriendly additions to the landscape, quite different from the total devastation seen in Dade County wherever residential development has supplanted rockland. Most houses on Big Pine Key are on lots left in a nearly natural state, and any alterations are usually in the form of horticultural enhancements using tropical vegetation that is tended and watered. This simulates the natural processes that take place in shallow limestone depressions described previously. Stacks of lumber, plywood window covers for hurricane protection, objects such as dumpsters and the spillage associated with them, and the piles of debris and trash that go hand-in-hand with residential communities in rural settings, while unattractive to human eyes, actually provide excellent microhabitat and cover for ringneck snakes. The reader may be surprised to learn that it is much easier to find *D. p. acricus*, one of America's rarest snakes, in a backyard garden than in pristine pine rockland.

Ecology and Natural History

Quite simply, there is very little known about this snake's natural history. Speculation regarding what could be learned from a detailed study of this taxon is provocative, because it lives in a vastly different environment than any of its congenerics. The environmental pressures on this snake are bound to be quite severe, the constraints on its activities seemingly cruel. One can reasonably suppose that *D. p. acricus* is an organism that persists under a very tenuous balance with its environment, barely eking out its continued survival through the generations. As Weaver and Christman (1978) urged, detailed natural history studies of this interesting and rare form are urgently needed.

Conservation

Conservation biologists often warn that when an organism is limited to one or a few tiny enclaves, it is extremely vulnerable to natural stochastic perturbations in its environment. Such small populations are easily extinguished. Without nearby populations to provide recruits to a dying counterpart or compensate for the loss of one population by their own continued flourishing, a single unpredicted event can cause the extinction of an entire taxon. This danger is so pervasive in the modern day, with so many plants and animals consigned to mere fragments of their historical distributions, that an entire facet of conservation biology—the management of captive "assurance colonies" in zoological parks—is based upon it.

On October 24, 2005, Hurricane Wilma raked past Big Pine Key as a Category 3 storm. The storm surge was 61–122 cm and inundated most of the island, as well as neighboring keys that harbored populations of Key Ringneck Snakes. Many areas remained submerged for days, completely saturating the ground with salt water and killing all vegetation in places, from ground cover to slash pine canopy. When the water finally receded and damage assessments were made, it became obvious that an ecological disaster had been wrought on Key Deer National Wildlife Refuge, as well as most of the unincorporated tracts of pine rockland on Big Pine and surrounding keys. Backyard oases were stripped of their lush plantings. Seawater had seeped into every fissure and cavity in the limestone, and inevitably, many ringneck snakes sheltering within were drowned. Most of the habitat features important to *Diadophis* were stripped away by water and wind and rendered uninhabitable for an extended period. The ringneck snakes that survived Hurricane Wilma's immediate destruction will face a very difficult time finding shelter and prey, and the future for this taxon looks very grim.

The effect of hurricane damage to natural habitat in the Big Pine Key rocklands contrasts with the scenario observed after Hurricane Ivan struck Oak Toad habitat along the Alabama coast. The toads were clearly not seriously impacted by the damage suffered by the forest they inhabited, judging by their conspicuous numbers seen in the months following the storm. Unlike the toads, the ringneck snakes on Big Pine Key had no other source population from which to draw new recruits. Hurricane Wilma pushed a deadly storm surge on top of the sum total of all extant individuals of *D. p. acricus*, whereas Hurricane Ivan merely impinged on the individual Oak Toads comprising one of a multitude of populations in the region. The toads, through the progeny of storm survivors and those from outlying populations, were able to quickly repopulate areas briefly decimated by flood and wind.

Summary

The Key Ringneck Snake is the sort of organism that would be well served by a breeding program and captive population management. Though native populations live primarily on protected and well-managed land, the tiny distribution and harsh environment of the Lower Key pine rocklands make

D. p. acricus very vulnerable. Consequently, it is listed as Threatened by the State of Florida. Hurricane Wilma was precisely the type of natural disaster feared by conservation biologists for the sake of organisms so precariously positioned. The population status of this snake has been dealt a hard blow, and whether it can recover is uncertain. If a captive assurance colony had been in place, the option of restarting decimated survivor populations with introduced captive-bred specimens would be available. Unfortunately, lack of interest in the taxon and the no doubt daunting challenges in captive husbandry have prevented such a program from being developed. If the Key Ringneck Snake survives Hurricane Wilma into its next generation, perhaps it will be seen as a second chance for conservationists to prepare for the next disaster that will surely befall this vulnerable chain of islands.

Rim Rock Crowned Snake
Tantilla oolitica

No other American species of reptile or amphibian has experienced a more thorough urbanization of the entirety of its habitat than the Rim Rock Crowned Snake. The transformation has been so complete that a present day reconnoiter of its historical range may lead one to wonder why *T. oolitica* is considered a pine woods species at all. It is simply beyond the memory of most people living today that Miami-Dade County was once dominated by a pine rockland, now only weakly suggested by the distinctive geological character of the ground. The other rockland endemic, the Key Ringneck Snake, has also experienced profound transformation of its haunts by a steadily growing human presence in south Florida, but appears to be tolerating the changes, as explained in the previous account. The Rim Rock Crowned Snake, in contrast, is more likely to be threatened by the intrusion because the urbanization is more uncompromising and engulfing. Whereas the ringneck snake can access natural areas through green spaces and corridors, the fragmented populations of *Tantilla* are completely isolated and cut off from each other, splintering the species into small discrete groups, a sure recipe for local extinctions.

The sorry state of most of south Florida's pine rockland is exemplified by *T. oolitica*'s humble type locality. At least as far back as 1955, when the holotype was collected by an unknown individual (passed on to the Florida State Museum by Dennis R. Paulson), the site was severely affected by the expanding development of suburban Miami. The specimen was found hiding beneath debris in a small vacant lot alongside SW 27th Avenue near the intersection of 24th Street. Based on the limestone surface features and scattered slash pines, the location had clearly once been a pine rockland completely subsumed by urban development. Here, cut off from the surviving environs of natural habitat and clinging to existence, one or perhaps several individuals of *T. oolitica* remained (plate 7.3).

Revisiting this site in 2004 revealed a scene probably very similar to the one in the 1950s. In all directions save one, there is no area even remotely suitable for a snake to survive, the only features being storefronts, sidewalks, and the constant bustle of vehicles and pedestrians along a busy city street. Off to one side, a vacant lot stands, quite possibly the same one where the holotype was found, and certainly very similar to it in most features. Here one could picture a tiny secretive snake still hiding, still persisting, feeding on the invertebrate fauna beneath the litter and refuse of a typical abandoned lot on the edge of a city. Just an acre (0.4 ha) of overgrown weeds, broken bottles, discarded rotting clothes, and litter. However, repeated searches, when every movable object was overturned during favorable periods following afternoon and evening rain showers, revealed no snakes and surprisingly few insects or potential invertebrate prey. The snake collected here 40 years earlier may well have been the last remnant of a once thriving population that inhabited a pine rockland long erased from this site. Scruffy, rocky pine scrub is not the sort of landscape that inherently causes city planners and commercial developers to pause and consider alternative paths, and these rocklands had no champion to call attention to them. The beauty and unique value of these areas were not fully appreciated when they were in their pristine condition, and no one knew at the time that they harbored a unique species of snake that contributed to the biological distinction of south Florida. By the time this rockland endemic was discovered, the human onslaught on their habitat was already underway.

Description

Although both the Florida and the Rim Rock Crowned Snakes are diminutive creatures, the Rim Rock Crowned Snake grows somewhat larger, with adults ranging 17.5–28 cm in length. Their tiny dimensions, secretive behavior, and sparse occurrence in scattered locales make *T. oolitica* an unnoticed organism even by people who are literally standing on top of it (plate 7.4).

The distinctive black head, the hallmark of these snakes, is variable in *T.*

oolitica. Many specimens have solid black heads and napes when viewed dorsally. In these, the black patch extends posteriorly as far as the third row of dorsal scales. Laterally, the cap extends raggedly to almost contact the margin of the ventral scutes in places. The black marking is disrupted by light pigment over the nape in some specimens, creating a light collar, or ring, around the top and sides of the neck. It was once thought that these two color patterns were correlated to geography, with solid black-headed snakes restricted to Dade County and collared specimens occurring only in the Keys. As additional specimens were found, it became apparent that, at least in the Upper Keys, *T. oolitica* exhibits both patterns (Porras and Wilson 1979). As this snake has been taken in barren sandy substrates as well as more mesic organic surfaces in subtropical hammocks adjoining rockland, it is possible that Telford's (1966) crypsis hypothesis for head pattern differences in *T. relicta* is applicable to *T. oolitica* as well. Since it is unlikely that significant numbers of Rim Rock Crowned Snakes can be observed in situ, this hypothesis will be difficult to test.

Other than their distinctive head pigmentation, Rim Rock Crowned Snakes are unmarked. The dorsum beyond the nape is light tan, often grading to a lighter and faintly pinker hue. The venter is paler than the dorsum, being some shade of very light tan or white, again with faint pinkish overtones subcaudally.

From Miami southward, no other species of *Tantilla* occurs and no other sympatric snake bears the slightest resemblance to *T. oolitica*, making identification straightforward. *Tantilla oolitica* is physically distinguished from its pine woods congeneric *T. relicta* by the presence of two enlarged basal hooks on each hemipene, in contrast with the single hooks seen on *T. relicta* (Telford 1966). *Tantilla oolitica*'s dissimilarity to its most geographically proximate congeneric and sharing of important characters with another species more remotely distributed from it (*T. coronata*) was addressed by Telford in his important 1966 paper. *Tantilla coronata* shares the two-spined character state of the hemipene with *T. oolitica*, yet the two species are separated by most of peninsular Florida, which is inhabited by *T. relicta*, a species to which *T. oolitica* bears less resemblance. Telford noted that several other species of reptiles endemic to south Florida and the Upper Keys have their closest relationship with taxa found north of the Florida peninsula rather than with more adjacently distributed species found further south. He explained this seeming paradox by pointing out that during Pleistocene interglacials, much of peninsular Florida existed only as islands in the prehistoric Okefenokee Sea. It is plausible that during the multiple risings and fallings of sea levels, land bridges may have connected south Florida and the Upper Keys with areas that are now northern Florida and Georgia. What would eventually become central peninsular Florida remained insular until sea levels began to fall to their present levels, creating the Florida we see today. This would explain the distinctive endemic fauna of the central Florida sand ridges and the close relationship between species in extreme north and south Florida.

Distribution Notes

Tantilla oolitica is restricted to the strand of oolitic limestone bedrock stretching from Miami to approximately the level of Homestead, Florida, and then continuing in broken fashion into the Upper and Middle Keys. A specimen was recently found on Big Pine Key, so it may also occur naturally in the limestone rockland of the Lower Keys. The species' primary refuge is within an assortment of 15 public and private parks that intermix with a scattering of vacant lots and residential properties that harbor a semblance of rockland or hammock habitat characteristics.

The distribution in Monroe County is poorly known, but a handful of specimens have been found there, first in Key Largo, and later from Grassy Key (Porras and Wilson 1979). Reports of *Tantilla* from Key West are intriguing, as they are for *Diadophis*. There are ample areas in the Lower Keys where *T. oolitica* would find suitable habitat. However, it should be noted that the paleogeography of the Upper and Lower Keys, like the history of peninsular Florida, is complicated by the waxing and waning of land bridges during a long period of varying sea levels. Even if the presence of an established population of Rim Rock Crowned Snakes on Key West or Big Pine Key is confirmed, the possibility of transplantation to these sites through the energetic movement of landscaping soils and plants from south Florida could not be discounted.

Habitat

The profound alterations that have been forced on the range of this species create a literature describing a wide variety of habitats, including vacant city lots, trash piles, and pastureland (Duellman and Schwartz 1958). However, these records are for specimens extracted from degraded sites and are misleading about the habitat preferred by the snake.

Although pine rockland is a xeric environment, the circumstances under which many Rim Rock Crowned Snakes have been found suggests a proclivity for mesic microenvironments. Most captures have been made in areas where sand overlies the limestone bed. Surface debris, whether natural features or human refuse, provides cover and protection from high temperatures and desiccation.

Subtropical hammocks are a more abundant feature of rockland in Dade and Monroe counties than similar habitats inhabited by *Diadophis* in the Lower Keys. *Tantilla oolitica* utilizes these damp refugia and lives beneath leaf litter and rotting logs (Porras and Wilson 1979) (plate 7.5).

Ecology and Natural History

Very little is known about how this species interacts with its natural environment. Like all members of the genus, it is extremely secretive and possibly fossorial. Many of the specimens that have come to light were collected on roads after heavy rains. As many areas where it still occurs are in parks and small reserves, some aspects of its ecology may be disrupted as individuals adapt to disturbed conditions. Observing such an animal in situ would be virtually impossible, so the best opportunity for gathering insights would be through the study of captives. Unfortunately, *T. oolitica* poses many challenges as a captive, and to date no sustained effort to culture the taxon ex situ has been made.

This is a pine woods resident that has an unusual relationship with fire. While mild burns on a small scale are a natural part of the environment, large or unusually hot burns, as can occur during times of drought or following a long lapse in smaller fires, may inflict severe mortality on local populations (plate 7.6).

No direct observations have been made on the reproductive biology of *T. oolitica*. It likely strays but little from the generic parameters of small clutches (1–3) of elongate eggs.

Conservation

The Rim Rock Crowned Snake is one of the rarest snakes in the United States. Given that the processes that have decimated its habitat are still active and unrestrained, the Rim Rock Crowned Snake is endangered. It nonetheless receives no federal protection through the Endangered Species Act. Although designated as Threatened by the State of Florida, this designation has little capacity to halt or hamper the outward expansion of development into remaining unsecured tracts of pine rockland and subtropical hammock. The die has already been cast for the extent of occurrence of *T. oolitica* into the future, this being the assortment of parks and reserves already fenced off from continued development. Good habitat and no doubt persistent populations of *T. oolitica* still survive in some of these parcels, but the refuges are merely fragments of a once much larger, interconnected south Florida landscape, and it is hard to conceive of the circumstances that would allow these rocklands to recover to an extent approaching their former dominance in this region. The ability of *T. oolitica* and other sensitive rockland flora and fauna to survive in these disjunct preserves is questionable; because of the nature and appearance of this humble snake, little effort is expected to be made on its behalf.

Summary

The current population status of *T. oolitica* offers a glimpse into the likely future of most fauna and flora of the southern pine woods. The persistence of species will soon be completely dependent on the establishment of reserves with sufficient area and connectivity to support natural ecologies. Species that by circumstance occupy regions where conservation foresight was lacking or stymied by opposing values will become extinct. Those fortunate enough to occupy lands expressly maintained for their value as natural areas may survive as stable, albeit shrunken, facsimiles of their ancestral populations.

There are several ways that such sanctuaries can come about. Most obvious but unfortunately rare is the scenario whereby the pine woods organism in question is the primary focus of a conservation initiative, and lands are preserved expressly for its security. Currently, no reserve exists with the primary intent of sheltering one of the endemic pine woods reptiles or amphibians. However, species such as the Rim Rock Crowned Snake are quite appropriate for such measures, since viable populations could exist on modest acreage, and it would not be impractical to secure a fine refuge for this snake, nor financially unobtainable, nor too taxing to manage for optimal conditions. Only the will is lacking, particularly among the general public, to exert effort and resources for such an obscure and humble creature.

Another mechanism by which a pine woods reptile or amphibian may have populations protected on refuge lands is when it rides on the conservation coattails of a more charismatic and marketable species. Such is the scenario that has maintained excellent habitat for the Louisiana Pine Snake in Fort Polk. Although the military base was not established with any intent to preserve wildlife, it has in fact done so admirably, chiefly by its intense program of prescribed burning and longleaf pine restoration and management driven by the federal mandate to provide recovery options for the Red-cockaded Woodpecker, an Endangered Species. While the Louisiana Pine Snake is not the intended recipient of the managerial decisions made to maintain upland longleaf habitat in the low-use portion of Fort Polk, both the Baird's Pocket Gopher and the pine snake benefit greatly from these actions.

The majority of pine woods herpetofauna populations situated on protected land are present as indirect recipients of good circumstances. The Striped Newts in Ocala National Forest, the Dusky Gopher Frogs in De Soto National Forest, the Flatwoods Salamanders living on Eglin Air Force Base, and numerous other examples are fortunate side effects of broad conservation initiatives that secured good habitat for the entire biological community in a particular location. These broadly encompassing managed habitats contribute to the conservation of plants and animals regardless of their inclusion or omission from state and federal endangered lists, and their breadth of coverage of pine woods herpetofauna is impressive. All pine woods endemics have populations located within public lands managed primarily as natural habitat, and there they have their best chance to survive into the distant future. Today, the only chance one has to see a Rim Rock Crowned Snake is on a public or private wildlife reserve. The days of finding this species in a suburban Miami backyard or under a piece of limestone at the side of a roadway spanning a rural zone between south Florida towns are gone. One day soon this will be the case for most of the species highlighted in these pages.

8

Other Pine Woods Herpetofauna

While it is true that virtually any reptile or amphibian occurring in the southeastern United States may on occasion be encountered in the pine woods (I have found Cottonmouths, *Agkistrodon piscivorus*, and Mud Snakes, *Farancia abacura*, in the most xeric of sandhill situations because of adjacent baygall wetlands), an assemblage of taxa is inextricably tied to this ecosystem. That fauna is the subject of this book. However, a faunal assemblage cannot be pigeonholed into a rigid category defined by habitat. There is a component of the herpetofauna that cannot be clearly placed among either the pine woods endemic subset or the species that are only rarely found in such situations. A unifying factor with most of these species is their predilection for open, sunny habitats. Such environments are fundamental to most southern pine woods ecosystems; hence the tendency to find these species here, but other associations also provide suitable conditions. Thus it is not accurate to consider these animals true endemics of the pinelands, and they fail the hypothetical test that if the pine woods were extirpated from the southern United States, it would directly cause the extinction of the following reptiles and amphibians. Since this book may find a significant portion of its readership among biologists who work in pine-dominated ecosystems, I have included brief overviews of a dozen reptile and amphibian taxa that, although not wholly restricted to such habitat, are encountered most often in these situations.

Ornate Chorus Frog
Pseudacris ornata

This gorgeous amphibian is rarely seen except by those who expend considerable effort to seek it out. It stays hidden for much of the year. On rare occasions, specimens are unearthed beneath centimeters of ground, or when large fallen objects are turned. As with most amphibians living in the pine woods, very little is known about the habits and ecology of *P. ornata* at times other than during the breeding season.

This is one of our truly beautiful frogs, marked as if painted with green, black, and red. The larva is small, reaching a maximum total length of 25 mm. It is an overall dark reddish brown creature, with indistinct smudges of yellow on the body, especially on the side of the head (Gregoire 2005) (plate 8.1).

The reproductive timing and habitat of *P. ornata* is similar to that of the Dusky Gopher Frog in many respects. The two species were historically spatially and temporally sympatric breeders in southwestern Alabama and southeastern Mississippi. Breeding coincides with periods of heavy rainfall, so the timing is variable, but generally occurs sometime during late fall through early spring. Mount (1975) briefly summarized the breeding habits in Alabama. Spawning occurs December through March. Preferred breeding sites are ephemeral ponds of shallow depth, with abundant emergent vegetation from where males call and upon which the eggs are usually attached. Clutches are fairly small; Mount referred to an account of 25 eggs. These transient pools are the kind once numerous in pine savanna, before the advent of widespread fire suppression policies, and such environments are quite rare today (plate 8.2).

Although savannas and flatwoods afford good habitat for this frog, it has also been found in natural meadows (Conant and Collins 1991), cypress ponds, and shallow Carolina Bays.

Gopher Tortoise
Gopherus polyphemus

The Gopher Tortoise, a sun-loving animal that favors well-drained open habitat, is a frequenter of some pine wood systems, but not restricted to them. Both ridge and savanna provide the warm, dry conditions and loose soils the tortoise prefers, but other types of scrubby associations, coastal dunes, and open fields are regularly utilized. Being an inhabitant of xeric sites, the Gopher Tortoise frequents pine ridge and savanna as a subset of this category. Thus, although often thought of in conjunction with pine woods, the Gopher Tortoise is not a true endemic (plate 8.3).

Many coastal environments provide good conditions for Gopher Tortoises, particularly areas behind the foredunes, where old dunes support scrub that may or may not include pine (Hartman 1978). Open, grassy situations, either natural prairie or manmade in the form of old fields, are sometimes occupied by *Gopherus* (Wilson et al. 1997). Although declining and rare in some areas, particularly near the western limits of its distribution, the relatively broad habitat parameters acceptable to this species result in its being present in all 67 Florida counties. It is also found in disjunct but extensive areas in southern Georgia, and in scattered locales in South Carolina, Alabama, and Mississippi (Wilson et al. 1997).

Though not wholly restricted to pine woods habitat, the tortoise impacts this ecosystem to a degree not approached by any other herpetofaunal element. The shelter afforded by its deep burrows is utilized by a cadre of animals, ranging from arthropods to mammals (Franz 1986). No other refuge in the pine woods provides such certain protection from heat and desiccation, and some species would be unable to survive in the more xeric sites without access to them. Gopher Tortoises influence floral understory structure through seed dispersal via their feces, which in turn affects the herpetofauna by shaping ground cover characteristics and creating food sources for particular invertebrates that some reptiles and amphibians feed upon. Grasses, which are a major component of ridge and savanna understory vegetation, are affected by the presence of *Gopherus*, as tortoises are likely the primary seed dispersers of native grasses in xeric pine habitat (Auffenberg 1969).

Gopher Tortoises require loose soil to excavate burrows. The structures can extend for 9 meters and lie 3 or more meters below the surface. The great impact of the tortoise on southern pine woods ecology is owed to the fact that a great variety of organisms utilize these burrows. Tortoise burrows provide a stable combination of moderate temperatures and comparatively high humidity in a landscape characterized by a paucity of moisture and decimating surface temperatures. For species such as the Dusky Gopher Frog, *Lithobates sevosus*, and Striped Newt, *Notophthalmus perstriatus*, few microhabitats as optimal as *Gopherus* burrows are available. Hardier taxa such as the Eastern Pine Snake, *Pituophis melanoleucus*, Eastern Indigo Snake, *Drymarchon corais couperi*, and Eastern Diamondback Rattlesnake, *Crotalus adamanteus*, also find tortoise burrows to be good transient shelters (Landers and Speake 1980), and favor them as sites to oviposit or give birth. Gopher Tortoises occupy a unique position in the herpetofauna of the pine woods in that while they are not dependent on pinelands, they play a crucial role in the life histories of endemic taxa wherever they do occupy such environments.

Gopher Tortoises shun deep pine forest with closed overstories because they require good sun exposure of the forest floor. Unable to climb and relegated to a small home range, Gopher Tortoises need an abundance of sunny basking sites where they can thermoregulate. As a result, pine habitats where fire has been suppressed are eventually abandoned.

Eastern, Slender, and Island Glass Lizards
Ophisaurus ventralis, O. attenuatus, and *O. compressus*

Of the four species of glass lizards native to the United States, only *Ophisaurus mimicus* is restricted to pine-dominated habitat. From central coastal South Carolina south and west to Apalachicola, which represents the majority of the distribution of *O. mimicus,* all three other species of *Ophisaurus*—the Eastern, *O. ventralis,* Slender, *O. attenuatus,* and Island, *O. compressus*—may be encountered as sympatrics. Across the entirety of the range of *O. mimicus,* the two most common and widespread species, *O. ventralis* and *O. attenuatus,* are also found. The Eastern, Slender, and Island Glass Lizards are partial to high, dry pine savannas, but not limited to them, so all four native species not only overlap in distribution but also often share the same specific localities and microhabitats. In coastal South Carolina, North Carolina, Georgia, and northern Florida, it is possible to lift an old board or sheet of corrugated tin and find all four glass lizard species lying underneath. Few unclaimed opportunities for the student of native lizard ecology are riper for delivery of fascinating insights than the study of the resource partitioning that allows this sympatry of four closely related and superficially similar congenerics (plate 8.4).

Species-specific habitat preferences broaden the range of suitable environments such that none can be considered endemic to pinelands. *Ophisaurus compressus* favors island habitats and coastal situations. Very much like the Gulf Coast Box Turtle, it is found in two adjacent but different associations—the barren dunes just beyond the reach of tidal wash, and the comparatively mesic pine savannas that commence slightly farther inland. *Ophisaurus attenuatus* is found in a variety of situations all sharing the characteristic of sparse mid- and overstory vegetation. Any relatively dry, sparsely treed habitat may support this species; thus such diverse environments as farmland, natural meadows, successional grassland and fields on abandoned pastureland as well as pine savanna are utilized. *Ophisaurus ventralis* is the most commonly encountered species. It prefers more mesic environments than the others. Consequently, pine flatwoods and the transitional zones with savanna are typical habitat, but any wooded or open field situation affording moisture may harbor the Eastern Glass Lizard.

Ophisaurus are encountered less frequently than they once were. For example, from 1977 to 1980 an average of four Slender Glass Lizards were brought to the Memphis Zoo each year by area residents who had captured the odd-looking creatures and wished to have them identified. Most of these specimens were found in Desoto and Panola counties in northern Mississippi just south of Memphis. However, since 1980, not a single glass lizard has been brought to the Zoo and this author is unaware of any reliable sightings from west Tennessee or northwest Mississippi since that year. Although possibly a coincidence, the glass lizard vanished from the area during the same year fire ants expanded their northern front, reaching the area in the late 1970s.

Six-lined Racerunner
Apidoscelis sexlineata

Sunny and dry—these are the two environmental characteristics associated with the habitat of the only native U.S. teiid found east of the Mississippi River. Pine ridge scrub is thus an especially suitable situation for the Six-lined Racerunner. This lizard is terrestrial, and usually spied during hot, sunny periods in the day after the ground has warmed sufficiently to its liking. These diurnal, fervidly active reptiles are usually the first herpetofauna noted when one enters territory they inhabit because of their habit of foraging over bare ground. Their jerky, erratic movements cause considerable rustling sounds when they traverse brushy vegetation. They also leave conspicuous physical signs of their presence, including distinctive tracks on sandy ground and numerous holes dug in their search for prey (plate 8.5).

The racerunner is well suited for the harsh conditions of coastal strand pine woods. Here the environment is continually restructured by the forces of wind and surf, a process ranging from imperceptibly slow shifting of dunes to periodic obliteration of whole scrub communities during tropical storms. *Apidoscelis* requires only barren, sunny tracts of sand, adapting well to major natural or manmade changes in its environment.

The racerunner is among the very few reptiles or amphibians one is likely to see fully exposed and active during hot summer middays in the pine woods. Most specimens are encountered in open, featureless locations, away from trees and other shading structures. Racerunners stand apart from most herpetofauna in the South by being physiologically adapted to be comfortable and efficient at an extremely high body temperature. The optimum body temperature for *A. sexlineata* while active is 40°C (104°F) (Mount 1975). This requires that racerunners stay fully exposed to the sun for much of the day and concentrate their activities between the hours of 1100 and 1500. This is quite the opposite from most diurnal reptiles, which tend to be bimodally active, in the morning and again in the late afternoon as the temperature begins to moderate.

The sun-loving racerunner is found in many habitats other than pine woods and uses fields of all kinds, both natural and manmade. It is not a species requiring wild, pristine areas to survive, and adapts to encroachment by humans, often exploiting the edges of commercial development along the Gulf coast. They may be seen foraging in sunny construction sites, using discarded boards, insulation, and tarpaper as shelter.

Racerunners lay 1–5 eggs during the summer, and females can produce a second clutch during the season (Mount 1975). Hatchlings are similar in appearance to adults, but exhibit light blue coloration on the tail and are overall more boldly patterned.

Southern Fence Lizard
Sceloporus undulatus undulatus

Fence lizards favor pine woods over deciduous forest because of the greater abundance of sunny basking sites. In hardwood forests, unbroken shade often dominates and the best opportunity for a diurnal lizard to thermoregulate in direct sun is along edges where the forest interfaces with clearings. To bask along such ecotones requires a lizard to expose itself to predators, especially raptors that frequent open fields. Natural pine habitat is more open canopied, even in its interior, than other forest types, and *Sceloporus* can find good basking sites while remaining relatively inconspicuous and out of plain view (plate 8.6).

Because of its preference for hot, sunny environments, *S. undulatus* is more abundant in scrub than in pine forest. However, given an open canopy, the species is likely to be seen in flatwoods, savanna, or ridge associations. Some tree cover is essential, however, as *S. undulatus* is arboreal and spends much of its time perched on the side of tree trunks. It typically basks within a meter or two of the ground, affording it the opportunity to prey on ground-dwelling insects as well as those it finds on the tree. When threatened it quickly flees upward and out of reach.

Eastern Indigo Snake
Drymarchon corais couperi

This, the largest of North American snakes, favors an interesting variety of habitats. The Eastern Indigo Snake's relationship to the pine woods is unique among reptiles and amphibians, complex, and multifaceted. Southern pine woods are an integral part of the natural history of some populations, but other populations are located in areas devoid of pine habitat. Indigos inhabit terrain at opposite ends of the hydric spectrum. With regard to pine woods, indigo snakes are found in ridge and flatwood situations, depending on the season, and less typically in savanna, which is surprising since that habitat often abuts ridges and flatwoods. The species is not restricted to pine woods, and other important habitats include wetlands, agricultural areas, and subtropical hardwood hammocks (plate 8.7).

The varied habitats used by eastern indigos are apparent at the population level regionally as well as temporally among individuals in many single populations. In Georgia, north Florida, and now vacated South Carolina sites, the taxon is nearly endemic to ridge habitat, where it is a commensal of the Gopher Tortoise (Diemer and Speake 1983). Where moist or wetland habitat is adjacent to these sandy slopes, Eastern Indigo Snakes move into them during the spring, feeding on reptiles, especially snakes, as well as frogs, fish, small mammals, and birds (Irwin et al. 2003). In the late fall the snakes return to the high, dry pine hills, where they seek refuge from cold temperatures in tortoise burrows. Indigos select tortoise burrows for winter refuge to the exclusion of almost all other refuge where the two species are sympatric (Diemer and Speake 1983). This annual movement between winter dormancy sites in sandhills and flatwood of wetland summer feeding sites is typical of the species in the northern portion of its range (plate 8.8).

In the southern half of the Florida peninsula, where the species is locally abundant (Lawler 1997), the habitats chosen do not exhibit such a temporal shift. In general, these southern populations are not closely associated with pine woods. Here, Eastern Indigo Snakes occur in low, wet habitat that resembles the summer feeding grounds of northern populations. In essence, the south Florida populations are continually engaged in activity in which northern populations are engaged only during the warm months. South Florida indigos remain active throughout the year. These snakes exploit the network of manmade drainage canals that criss-cross the land in this region, as well as natural wetlands. Well-irrigated agricultural areas, including citrus groves and sugar cane fields, frequently harbor indigos and further exemplify the species' ability to adjust to some forms of human alteration of habitat. At the southern limit of its distribution, the species again exhibits adaptability to local habitat availability by frequenting subtropical hardwood hammocks and venturing into pine rockland in the Keys. The Eastern Indigo Snake embodies a unique variation on pine woods endemism in that only a segment of the global population can be so characterized, and in some of these cases, only during specific times of the year.

Eastern Coachwhip
Masticophis flagellum flagellum

These large, sleek, high-strung snakes are strictly diurnal, favoring the hotter hours of the day, the sunnier the better. In order to capture their wary lizard, snake, and warm-blooded diurnal prey, coachwhips must be sharp-eyed and alert to tactile, chemosensory, and visual stimuli. Coachwhip eyes positively sizzle with a fierce energy. Their skill in avoiding close human approach suggests that their vision is acute and crucial in providing cues. Eastern Coachwhips avidly avoid any contact with humans by slipping in the opposite direction at the first sign of danger, but they will defend themselves enthusiastically if their escape route is cut off (plate 8.9).

This snake is ubiquitous throughout most of the southern pinelands. Coachwhips favor open habitats where their restless daytime prowling is fueled by warm sunny substrates and their visually oriented predatory tactics aren't hampered by dense underbrush. They are a sun-loving species possessing comparatively high tolerance for hot summertime temperatures, and are not at their most efficient and alert in shady woodland. Consequently, the broken overstory and abundant sun-drenched clearings typical of many pineland associations are much to the coachwhip's liking. However, because similar conditions are provided by numerous other situations, such as livestock pastures, road and utility line right-of-ways, crop fields, and residential areas, the Eastern Coachwhip is not a symbolic species of the pine woods (plate 8.10).

Eastern Coachwhips are often the most conspicuous snake in the pine woods. Their obvious presence is in part due to their size—coachwhips are one of the largest snakes in the United States, known to reach 2.6 m in length (Conant and Collins 1991). Their bold nature also contributes to their perceived abundance. However, their numbers may truly dominate the snake fauna of a particular region, as a snake sampling study in Louisiana has revealed (Reichling, unpubl. data).

Coachwhips have an arsenal of effective defensive behaviors, ranging from meek to fiery. The first line of defense is to flee with amazing speed and focus. Because they are very alert and visually oriented, they become aware of an approaching threat sooner than other snake species; thus they often escape before they are noticed. If detected and pursued, they frequently ascend trees with the same graceful agility they display on the ground. The coachwhip's choice of arboreal refuge is often a small or isolated bush, and one of the easiest ways to capture one is to maintain a close eye when it retreats and corner it in a tree where it can be seized by hand or tongs.

When cornered or restrained, coachwhips first revert to fierce striking. One thus aroused will shake its tail, rear up in stacked S-shaped coils so that the head is high above the ground, and strike with a rapid succession of accurate, open-mouthed lunges. Large specimens can deliver a bite painful enough to distract a threatening human long enough to allow the snake to slip away.

If these tactics fail, Eastern Coachwhips feign death. Poking, prodding, or sometimes merely picking up an enraged coachwhip will trigger the snake to assume a compact coil and become limp. The head is spread slightly at the angle of the jaws, making it appear larger and more triangular, and the neck is held stiffly crooked so the snout is pointing toward the ground. The effectiveness of this behavior becomes apparent when watching encounters between coachwhips and dogs. Turning to fight enrages most dogs and would ultimately lead to injury or death for the snake. Death feigning elicits pawing or nipping, but dogs usually quickly lose interest in a limp snake and after a few final sniffs, leave the snake without serious injury.

Eastern Hognose Snake
Heterodon platirhinos

Like the Bluetail Mole Skink, Florida Worm Lizard, and Short-tailed Snake, the Eastern Hognose Snake is one of numerous species that finds suitable habitat in the pine woods by virtue of its sandy, friable substrate. *Heterodon* is well adapted for burrowing, and does so when seeking shelter or prey. Its enlarged and upturned rostral scale is efficient at displacing soil and assisting the snake while probing below the surface. Loose, sandy soil is not the exclusive provenance of the pine woods, and hognose snakes may turn up anywhere throughout much of the southern and eastern United States wherever soils are conducive for burrowing.

The common features of most Eastern Hognose Snake locales, in addition to the sandy substrate, are open sunny exposures and dry terrain (Wright and Wright 1957). In addition to pine woods, Eastern Hognose Snakes are found in agricultural settings where old fields, tilled earth, and numerous edges dividing clearings and windrows combine to provide favored conditions (plate 8.11).

Although *H. platirhinos* and the pine woods endemic *H. simus* are morphologically very similar, there are significant distinctions in ecology and behavior. *Heterodon simus* is restricted to xeric pine woods. The basis for this habitat specificity and the reason it is not shared by *H. platirhinos* are unclear. Behaviorally, *H. platirhinos* is much more prone to feign death following an unsuccessful defensive sham than *H. simus*. *Heterodon simus* appears more inclined to bask, as anecdotal accounts by field herpetologists repeatedly mention specimens found lying coiled and still in partial shade during the daytime (Justin Collins, pers. comm.). I have found numerous Eastern Hognose Snakes under cover or in the open and on the move, but never resting in repose and obviously basking in the open.

Scarlet Kingsnake
Lampropeltis triangulum elapsoides

The Scarlet Kingsnake is found in a variety of habitats throughout the Deep South, but within the pine woods it is often found in a very characteristic and distinctive situation. During the warmer months of the year Scarlet Kingsnakes occupy the landscape across a broad range of microhabitats. In the spring, summer, and fall, specimens may be found beneath stones, trash piles, and other human debris, under or within fallen decaying logs, or out prowling at night during rainy weather on rural roadways. During the winter months, when many other southern reptiles are difficult to find, Scarlet Kingsnakes can be reliably discovered hibernating beneath the loose bark of dead pine trees in savanna and flatwoods. The still standing trunks of dead trees, particularly those set in wet or partially flooded ground, when decay has caused the loss of branches and upper trunk, leaving only a short column of pulpy, dead wood, are a classic setting for Scarlet Kingsnakes in winter dormancy. The bark on such pine trunks, though still intact, is often separated from the wood beneath, creating a tight, moist space perfect for sheltering small invertebrates, amphibians, and reptiles. Although other pine woods reptiles are occasionally discovered under loose pine bark during the winter months, the Scarlet Kingsnake seems to utilize this microenvironment preferentially. In fact, it is not unusual to find several specimens beneath a single sheet of loose bark (plate 8.12).

Unfortunately, this interesting behavior leaves the snakes vulnerable to overcollection and microhabitat destruction. No pine woods reptile or amphibian should be removed from the wild unless a critically weighed decision has been made beforehand and the collection found to be well justified. Collection of the present species is especially difficult to do in deference to a good conservation ethic. Collecting winter-dormant *L. t. elapsoides* requires destroying the outer layers of a rotting log. Not only are individual snakes removed from a local population, but an important environmental resource is destroyed for any other specimens residing nearby. Winter collections of Scarlet Kingsnakes should be noted as being very destructive to the environment.

Eastern Diamondback Rattlesnake
Crotalus adamanteus

In its ability to command awe and respect, the Eastern Diamondback Rattlesnake has no peer. No other snake carries itself with such dignity when disturbed. Although not the largest of the world's venomous snakes, it distinguishes itself by its calm and alert response to close approach, whereas even the largest King Cobras, *Ophiophagus hannah*, Bushmasters, *Lachesis muta*, Taipans, *Oxyuranus* spp., and other large venomous snakes react with a mixture of frantic offense and panicked retreat. Most diamondbacks, when discovered and roused, respond by lifting their head slightly, locking their gaze directly on the potential danger, rattling briefly, and either holding their ground or moving off very slowly, never breaking their measuring gaze leveled at the intruder. Frantic, premature strikes or break-and-run escapes are not part of the diamondback's defense repertoire. Diamondbacks are courageous, confident, and dignified reptiles; measured and restrained in their offense, but capable of inflicting horrendous injury if provoked beyond the limits of their tolerance (plate 8.13).

Eastern diamondbacks are a characteristic presence in scrub habitat. They are less likely to be found in closed canopy, well-shaded pine woods situations. The sparsely canopied pine-palmetto flatwoods of Florida and the coastal lowland of South Carolina and Georgia are ideal habitats for this magnificent pit viper, but it also utilizes coastal dune scrub, marshy areas, and deciduous woodland (plate 8.14).

The diamondback has become scarce, or been extirpated, from many areas. Being large and not especially secretive, it is often encountered in areas where it is abundant. Being truly dangerous and feared, it bears the greatest burden of humankind's intolerance of snakes. Even some naturalists and outdoors men and women, lovers of nature or sports enthusiasts who generally follow a "live and let live" policy toward snakes, still find it hard to let a diamondback live in peace if encountered. It is the chief target of several "rattlesnake roundup" events that take place in the South each year, where snakes are tortured, exploited, and slaughtered under spurious justification. Eastern Diamondback Rattlesnakes are now rarely seen in Louisiana, and are localized in occurrence in Mississippi and Alabama. Large populations persist in South Carolina and Georgia. The snake's stronghold is in Florida, a state with which it is synonymous and where professional sports teams take its name in honor, but this vaunted profile likely affords them little mercy.

Epilogue

In Charles Dickens' classic, *A Christmas Carol*, Ebenezer Scrooge is visited on Christmas Eve by three ghosts who by turns show him the past, present, and future of the people whose lives his callous soul has affected. Scrooge's most frightening visitation is by the Ghost of Christmas Future, but he changes his ways after he glimpses how his uncaring actions will damage the lives of those around him. For Scrooge, it is not too late to change course, but what of our prospects for changing the way we treat the southern pine woods?

A look toward the likely future of the pine woods is indeed frightening. The predominant attitude toward these habitats for the past 100 years is that they are a resource trove to be plundered and used in whatever way we find pleasing, without regard for the flora and fauna. Timber is cleared, and then natural processes are perverted to accelerate the efficient production of more wood. Precious scrubs are razed for homes and commerce. Wetlands are destroyed. No less cruelly than Scrooge exploited his employees, we take what we value from the pine woods while decimating what treasure they really hold—plants and animals beautifully shaped by these lands and found nowhere else on Earth. Like Scrooge's Christmas haunting, today is surely our last opportunity to change course by refocusing our conservation ethic and raising southern pine woods conservation to the high priority it deserves.

For this book's youngest generation of readers, those who are in college or about to enter it, and who are embarking on careers in conservation biology or herpetology, I am looking to a day 30 or 40 years into the future. This is a day that will be shaped by the course you choose. Now is the time to realize how dire is the hope that functioning pine woods ecosystems will still exist in 2040, if current trends persist. In my 30 years of working in these landscapes, I have witnessed tremendous losses of habitat, profoundly negative shifts in management practices, and increased rarity of many species. Without visionary action, the next 40 years promise more of the same. What happens—or fails to happen—during these next four crucial decades will determine whether the species accounts and habitat descriptions in the preceding pages are still relevant and accurate, or have become an historical record of times and places lost to future generations.

In the near future, all semblance of sound pine woods habitat will be found in a few national forests and military bases. Parcels currently under private control are too small to maintain a complete fauna over generations; and there will be little opportunity to expand on these holdings. As for commercial timberlands, all signs indicate that the entirety of habitat now in such places will be eliminated over the next 30 years. When large boards and heavy lumber were still profitable, timber companies could manage land as true forests and still function as businesses. Today, timber companies are responding to depressed markets for domestic lumber by converting production to pulpwood or selling off their forests. Already many important refuges within commercial operations have been irreparably altered or divided into smaller parcels and sold to other business interests.

I fear that pine rocklands will no longer exist in south Florida by 2040. This habitat survives tenuously as an array of small, scattered preserves across the historical distribution. Most of the examples are illustrative of the appearance of rockland but lack the ecological layers that should lie beneath the visible surface. All are so severely impinged on all sides by fierce development that the ecosystems they contain will collapse. The more extensive rockland within Key Deer National Refuge offers some hope of persistence, but anyone who has seen the recent influx of commercial development into tiny Big Pine Key has reason to be very pessimistic. Along with a host of other endemic plants and animals, the Key Ringneck Snake and Rim Rock Crowned Snake will be made extinct by loss of pine rockland. Neither of these taxa has or is ever likely to have a captive breeding program in response to such a catastrophe.

Current trends predict the extinction of at least three pine woods taxa, at least in the wild. Clearly, and despite the strong efforts of conservationists, the Dusky Gopher Frog is headed for extinction. It teeters on the brink with perhaps fewer than 50 frogs extant at a given time. Just about everything that could be bad for this frog has materialized all at once. For this species, the pressures exerted by devastating disease, habitat degradation, genetic bottleneck, and aggressive development of surrounding land seem insurmountable. The Louisiana Pine Snake will finally fulfill the predictions that herpetologists have been making since the 1930s and cease to exist in Louisiana and Texas. The pine snake's only real strongholds, within commercial timberland, are being rapidly lost, and without commercial incentive or federal protection, there is nothing that seems to be able to stop it. The Apalachicola Lowland Kingsnake, barely recognized by biologists as the unique form it is because it has been formally described only recently, is also destined for oblivion in the absence of a strong rescue effort, but none is perceivable on the horizon. Its habitat is undergoing dramatic changes, commenced only a few years ago, now that this last undeveloped Florida coastline is squarely in the aim of land developers.

At some point along the steady decline of the southern pine woods, a time will come when there is little sense in preserving them. In other words, a conservation cause can be lost before the last tree is felled or the last survivor of a species perishes. This is the reason it sometimes seems that a species becomes critically endangered very suddenly. Pressures build with the increase of threatening factors, and the system, whether a biological community or a single population, becomes very unstable and fragile—though it appears unchanged—until the proverbial "last straw" is added, whereupon the entity rapidly enters a state of no return. When pine woods contract to the point that only a few small fragments are left here and there in the southern United States, and whole subdivisions have been completely lost, a coldly objective conservationist might reasonably suggest that, like a medical triage, hope of survival of southern pine woods should be abandoned for other issues that offer more promise of success. With the loss of a half dozen or so herpetofaunal species and perhaps complete biozones like the Trans-Mississippi sandhills, pine rockland, and Apalachicola wetlands—all prime candidates for that fate—throwing in the towel might be considered much sooner than many people expect.

Appreciation and concern for the southern pine woods will be relegated to arcane status by the domination of global environmental issues in the public consciousness. There is no question that we must find clean, safe energy sources, promote environmentally friendly practices, and prevent manmade global warming, and it is good that these topics are discussed as important issues by politicians and business leaders and are becoming household words and known to elementary school children. I fear, however, that more regional environmental concerns are losing their voice. There is only so much attention politicians will devote to environmental stewardship, only limited space on the agendas of conservation action plans, and only a finite amount of money and personnel to be devoted to tackling such problems. I am concerned that the increasing emphasis on global issues will make initiatives to restore pine rockland or manage habitat for the Flatwoods Salamander seem trivial.

Like Ebenezer Scrooge at the end of his visitation by the Ghost of Christmas Future, we can ask, "Is this the future of the southern pine woods that *will* be, or the future that *could* be if we do nothing to change it?" Do we have any power to alter the likely sad course of events that are decimating the herpetofauna of the pine woods? What can each of us do?

The loss of privately owned pine habitats will be a terrible blow to pine woods herpetofauna. To stem the decline, there must be incentives to protect the land and manage it for natural habitat preservation. The federally managed national forests are much more stable and dependable with regard to perpetuating pine woods ecosystems primarily because their tax-funded management offers the luxury of not needing to pay their own way. Concerning commercial timberlands, they harbor some exceptionally precious pine woods yet the owners have no compelling reason to preserve them and every reason to exploit them. I have no doubt that many land managers of commercial tracts love the land and wildlife and are fully aware of the effect of their operations, but reality dictates actions that foster the growth and harvest of forest products at the expense of other interests. To expect any timber company to do otherwise requires bargaining leverage and incentives, both of which are rarely available today. Biologists need to arm themselves with the tools to speak the language of business when they sit at the discussion table with businessmen.

The four most imperiled pine wood endemics currently lacking such protection—the Key Ringneck Snake, Rim Rock Crowned Snake, Louisiana Pine Snake, and Apalachicola Lowland Kingsnake—need federal Endangered or Threatened Species listings to be implemented as soon as possible. The "candidate" status for listing some of these taxa is a state of limbo that impedes action by giving the appearance that action is imminent. Each of these species needs champions who will petition for federal listing and stay actively involved in that process. Once this is accomplished, the power of the Endangered Species Act can be brought to bear to form recovery plans and critical habitat designations, and funding to implement interventions. For those students of conservation biology who are including environmental law and politics in their curricula, these would be worthy causes. Solid protection of these taxa would also protect a host of rare and endemic plant and animal sympatrics.

Somehow we must go backward in time to when prescribed burning was standard operating procedure on all managed pinelands. To be effective, fire must visit a site every three to five years, be sufficiently hot to destroy hardy undergrowth and sweep the forest clean, and take place during the summer growing season when plants are most vulnerable. Such burns are virtually nonexistent in the present time, but as recently as 40 years ago they were ubiquitous. Today's prescribed burn in the South is a wintertime event, ignited on days when moisture and temperature are such that the weak fire merely singes the top layer of pine straw and has little or no suppressive effect on understory growth. In many instances these anemic fires actually accelerate underbrush development by stimulating new and denser regrowth from roots and stems that were damaged but not killed by the flames. Conservation biologists working in the southern United States should seize every opportunity to champion and cajole to bring summer fire back into the picture for the pine woods.

The reader will perceive that I am pessimistic about the future of the southern pine woods. Indeed, I think this book is likely to be viewed in the future as a record of the swan song of the amazing creatures that reside in flatwoods, ridge, and savanna. The means to restore and preserve these biological treasures are clear and simple in theory, but myriad practical problems block their implementation to any significant extent. For these forests and scrub, the clock hands are truly a few minutes before midnight, after which it will be too late. When viewed from a global perspective, the clarity needed to see the importance of these landscapes becomes fogged by the multitude of other problems visible to those with an environmental conscience. There are, however, enough conservation biologists to bring their talents and passions together to stop the incessant erosion of the most distinctive natural landscape of the American South. I hope that the herpetologists of 40 years hence will be able to visit the sites of the present-day pine woods and find each of the species described in this book still acting out their lives in the settings I've described. Those who ensure by their works that this is the version of the future that unfolds will have established a great legacy of conservation as their mark.

Bibliography

Allen, E. R. 1939. "Habits of *Rhadinaea flavilata*." *Copeia*, 1939: 175.

Allen, E. R., and W. T. Neill. 1953. "The Short-tailed Snake." *Florida Wildlife*, 6(11): 8–9.

American Association of Zoos and Aquariums. 2005. "AZA Conservation Programs." In *The 2005 AZA Membership Directory*, edited by Jane Ballentine, 271–304. Silver Spring, Md.: American Association of Zoos and Aquariums.

Anderson, K., and P. E. Moler. 1986. "Natural hybrids of the Pine Barrens Treefrog, *Hyla andersoni*, with *H. cinerea* and *H. femoralis* (Anura: Hylidae): morphological and chromosomal evidence." *Copeia*, 1986: 70–76.

Arnold, S. J. 1977. "The evolution of courtship behavior in New World salamanders with some comments on Old World salamandrids." In *The Reproductive Biology of Amphibians*, edited by D. H. Taylor and S. I. Guttman, 141–83. New York: Plenum Press.

Ashton, R. E., and P. S. Ashton. 1981. *Handbook of Reptiles and Amphibians of Florida, Part One: The Snakes*. 2nd ed. Miami: Windward. 176 pp.

Auffenberg, W. 1963. "The fossil snakes of Florida." *Tulane Studies in Zoology*, 10: 131–216.

———. 1969. *Tortoise Behavior and Survival: Patterns of Life Series*. Chicago: Rand McNally.

Babbit, K. J., and G. W. Tanner. 1997. "Effective management for frogs and toads on Florida's ranches." WEC16, Wildlife Ecology and Conservation Department, Florida Cooperative Extension Service, Institute of Food and Agricultural Sciences, University of Florida, 1–6.

Babbitt, L. H. 1951. "Courtship and mating of *Eumeces egregius*." *Copeia*, 1951: 79.

Beane, J. C. 1990. "*Rhadinaea flavilata*: geographic distribution." *Herpetological Review*, 21: 41–42.

Beane, J. C., T. J. Thorp, and D. A. Jackan. 1998. "*Heterodon simus*. Diet. Natural history notes." *Herpetological Review*, 29: 45.

Bishop, S. C. 1941. "Notes on salamanders with descriptions of several new forms." *Occasional Papers of the Museum of Zoology, University of Michigan*, 451: 1–21.

———. 1943. *Handbook of Salamanders. The Salamanders of the United States, of Canada, and of Lower California*. Ithaca, N.Y.: Comstock.

Blaney, R. M. 1977. "Systematics of the Common Kingsnake, *Lampropeltis getulus* (Linnaeus)." *Tulane Studies in Zoology and Botany*, 19: 47–104.

Boyer, W. D. 1979. "Regenerating the natural longleaf pine forest." *Journal of Forestry*, 77: 572–75.

Brenneman, L., and W. F. Tanner. 1958. "Possible abandoned barrier islands in panhandle Florida." *Florida Journal of Sedementology and Petrology*, 28: 342–44.

Brown, A. E. 1890. "On a new genus of Colubridae from Florida." *Proceedings of the Academy of Natural Sciences of Philadelphia*, 42: 199–200.

Bullard, A. J. 1965. "Additional records of the treefrog *Hyla andersoni* from the coastal plain of North Carolina." *Herpetologica*, 21: 154–55.

Burbrink, F. T. 2002. "Phylogeographic analysis of the cornsnake (*Elaphe guttata*) complex as inferred from maximum likelihood and Bayesian analyses." *Molecular Phylogenetics and Evolution*, 25: 465–76.

Burger, J., and R. T. Zappalorti. 1991. "Nesting behavior of pine snakes (*Pituophis m. melanoleucus*) in the New Jersey Pine Barrens." *Journal of Herpetology*, 25: 152–60.

Campbell, H. W. 1978. "Short-tailed Snake." In *Rare and Endangered Biota of Florida. Vol. 3. Reptiles and Amphibians*, edited by R. W. McDiarmid. Gainesville: University Presses of Florida.

Campbell, H. W., and W. H. Stickel. 1939. "Notes on the Yellow-lipped Snake." *Copeia*, 1939: 105.

Carr, A. F., Jr. 1934. "Notes on the habits of the Short-tailed Snake." *Copeia*, 1934: 138–39.

———. 1940. "A contribution to the herpetology of Florida." *University of Florida Publications (Biological Science)*, 3: 1–118.

———. 1949. "Notes on the eggs and the young of the lizard *Rhineura floridana*." *Copeia*, 1949: 77.

Christman, S. P. 1970. "*Hyla andersoni* in Florida." *Quarterly Journal of the Florida Academy of Science*, 33: 80.

Christman, S. P., and L. R. Franz. 1973. "Feeding habits of the Striped Newt, *Notophthalmus perstriatus*." *Journal of Herpetology*, 7: 133–35.

Cliburn, J. W. 1980. "Pine snakes seen by few." *Mississippi Outdoors*, 43: 23.

Collins, J. T., and T. W. Taggart. 2002. *Standard Common Names and Current Scientific Names for North American Amphibians, Turtles, Reptiles & Crocodilians*. Lawrence, Kans.: The Center for North American Herpetology. iv + 44 pp.

Compton, V. 2000. "Development of conservation strategies and projects." Final Report. *The Gulf Coastal Plain Ecosystem Partnership*, report number A613773, contract number DAMD17-98-2-8015. 274 pp.

Conant, R. 1956. "A review of two rare pine snakes from the gulf coastal plain." *American Museum of Natural History Novitates*, 1781: 1–31.

———. 1958. *A Field Guide to the Reptiles and Amphibians of the United States and Canada East of the 100th Meridian*. Boston: Houghton Mifflin.

Conant, R., and J. T. Collins. 1991. *A Field Guide to Reptiles and Amphibians. Eastern and Central North America*. Boston: Houghton Mifflin.

Desmond, D. 2001. "Species: southern hognose (*Heterodon simus*)." Herp Tech Associates: www.herptech.org, 1–4.

De Vosjoli, P. 1994. "Herpetoculture in a changing world." *Vivarium*, 6: 10–11, 36–37.

Diemer, J. E., and D. W. Speake. 1983. "The distribution of the Eastern Indigo Snake, *Drymarchon corais couperi*, in Georgia." *Journal of Herpetology*, 17: 256–264.

Ditmars, R. L. 1939. *A Field Book of North American Snakes*. Garden City, New York: Doubleday. xii + 305 pp.

Dodd, K., Jr. 1993. "Cost of living in an unpredictable environment: the ecology of Striped Newts *Notophthalmus perstriatus* during a prolonged drought." *Copeia* 1993: 605–14.

Dodd, K., Jr., and B. G. Charest. 1988. "The herpetofaunal community of temporary ponds in north Florida sandhills: species composition, temporal use, and management implications." In *Management of Amphibians, Reptiles, and Small Mammals in North America*, edited by R. C. Szaro, K. E. Severson, and D. R. Patton, 87–97. General Technical Report RM-166. Fort Collins, Colo.: USDA Forest Service, Rocky Mountain Forest and Range Experimental Station.

Dodd, K., Jr., and L. V. LaClaire. 1995. "Biogeography and status of the Striped Newt (*Notophthalmus perstriatus*) in Georgia, USA." *Herpetological Natural History*, 3: 37–46.

Donoghue, J. T. 1989. "Sedimentary environments of the inner continental shelf, northeastern Gulf of Mexico." *Transactions of the Gulf Coast Geological Society*, 39: 355–64.

Dowling, H. G., and L. R. Maxson. 1990. "Genetic and taxonomic relations of the Short-tailed Snakes, genus *Stilosoma*." *Journal of Zoology*, 221: 77–85.

Duellman, W. E., Schwartz, A. 1958. "Amphibians and reptiles of southern Florida." *Bulletin of the Florida State Museum, Biological Sciences*, 3: 181–324.

Dunn, E. R. 1940. "The races of *Ambystoma tigrinum*." *Copeia*, 1940: 129–30.

Duran, C. M. 1998a. "A radio-telemetry study of the Black Pine Snake (*Pituophis melanoleucus lodingi* Blanchard) on the Camp Shelby training site Camp Shelby, Mississippi." Unpublished. Final Report to the Mississippi Natural Heritage Program and the Mississippi Army National Guard.

———. 1998b. "Status of the Black Pine Snake (*Pituophis melanoleucus lodingi* Blanchard)." Unpublished report submitted to U.S. Fish and Wildlife Service, Jackson, Miss. 32 pp.

———. 2000. "Quantitative analysis of the status of the Black Pine Snake (*Pituophis melanoleucus lodingi*)." Unpublished report submitted to U.S. Fish and Wildlife Service, Jackson, Miss. 15 pp. + appendixes.

Earley, L. S. 2004. *Looking for Longleaf: The Fall and Rise of an American Forest*. Chapel Hill and London: The University of North Carolina Press. 322 pp.

Edgren, R.A. 1955. "The natural history of the hog-nosed snakes, genus *Heterodon*: a review." *Herpetologica*, 11: 105–17.

Edwards, G. B. 1994. "McCrone's Burrowing Wolf Spider." In *Rare and Endangered Biota of Florida*. Vol. IV. *Invertebrates*, edited by N. Deyrup and R. Franz, 232–33. Gainesville: University Press of Florida.

Enge, K. M. 1994. "*Rhadinaea flavilata*: geographic distribution." *Herpetological Review*, 25: 168.

———. 1997. "Habitat occurrence of Florida's native amphibians and reptiles." Technical Report No. 16, Tallahassee: Florida Game and Freshwater Fish Commission.

Ernst, C. H. 1981. "Courtship behavior of male *Terrapene carolina major* (Reptilia, Testudines, Emydidae)." *Herpetological Review*, 12: 7–8.

Franz, R. 1986. "*Gopherus polyphemus*. Burrow commensals." *Herpetological Review*, 17: 64.

——. 1992. "Florida pine snake." In *Rare and Endangered Biota of Florida*. Vol. 3. *Amphibians and Reptiles*, edited by Paul Moler, 254–58. Gainesville: University Presses of Florida.

Frost, C. C. 1993. "Four centuries of changing landscape patterns in the longleaf pine ecosystem." In *Proceedings of the 18th Tall Timbers Fire Ecology Conference*, 18: 17–43.

Gans, C. 1967. "*Rhineura floridana* (Baird), Florida Worm Lizard." *Catalogue of American Amphibians and Reptiles*, 43.1–43.2.

Garman, S. W. 1883. "Notes on certain reptiles from Florida and Brazil." *Science Observer*, 4: 47–48.

Gerald, G. W. 2005. "Natural history notes: *Ophisaurus attenuatus longicaudus*. Clutch size and brooding." *Herpetological Review*, 36: 181–82.

Gibbons, J. W., and R. D. Semlitsch. 1987. "Activity patterns." In *Snake Ecology and Evolutionary Biology*, edited by R. A. Seigel, J. T. Collins, and S. S. Novak, 396–421. New York: Macmillan. xiv + 529 pp.

Godwin, J. C. 1992. "Geographic distribution: *Rhadinaea flavilata* (Pine Woods Snake)." *Herpetological Review*, 23: 92.

Goin, C. J. 1947. "A note on the food of *Heterodon simus*." *Copeia*, 1947: 275.

Goodyear, C. P. 1971. "Y-axis orientation of the Oak Toad, *Bufo quercicus*." *Herpetologica*, 27: 320–23.

Gosner, K. L., and I. H. Black. 1957. "The effects of acidity on the development and hatching of New Jersey frogs." *Ecology*, 38: 256–62.

Gottlieb, J. A. 1984. "*Rhadinaea flavilata*: geographic distribution." *Herpetological Review*, 15: 21.

Greenberg, C. H., A. Storfer, G. W. Tanner, and S. G. Mech. 2003. *Amphibians Using Isolated, Ephemeral Ponds in Florida Longleaf Pine Uplands: Population Dynamics and Assessment of Monitoring Methodologies. Final Report*. Bureau of Wildlife Diversity Conservation, Florida Fish and Wildlife Conservation Commission, Tallahassee. 38 pp.

Greenberg, C. H., and G. W. Tanner. 2005. "Spatial and temporal ecology of Oak Toads (*Bufo quercicus*) on a Florida landscape." *Herpetologica*, 61: 422–34.

Gregoire, D. R. 2005. *Tadpoles of the Southeastern United States Coastal Plain*. Southeast Amphibian Research and Monitoring Initiative, U.S. Geological Survey. 60 pp.

Griswold, W. G. IV. 2000. "The Southern Hognose Snake: all you ever wanted to know." *Reptiles*, 8: 32–45.

Guidry, E. V. 1953. "Herpetological notes from southeastern Texas." *Herpetologica*, 9: 49–56.

Hamilton, W. J., Jr. 1955. "Notes on the ecology of the Oak Toad in Florida." *Herpetologica*, 11: 205–10.

Hamilton, W. J., Jr., and J. A. Pollack. 1958. "Notes on the life history of the red-tailed skink." *Herpetologica*, 14: 25–28.

Hardy, J. D., Jr. 1952. "A concentration of juvenile Spotted Salamanders *Ambystoma maculatum* (Shaw)." *Copeia*, 1952: 181–82.

——. 1969. "Reproductive activity, growth, and movements of *Ambystoma mabeei* in North Carolina." *Bulletin of the Maryland Herpetological Society*, 5: 65–76.

Hardy, J. D., Jr., and J. D. Anderson. 1970. "*Ambystoma mabeei*." *Catalogue of American Amphibians and Reptiles*, 81.1–81.2.

Hardy, J. D., Jr., and J. Olman. 1974. "Restriction of the range of the frosted salamander, *Ambystoma cingulatum* based on a comparison of the larvae of *Ambystoma cingulatum* and *Ambystoma mabeei*." *Herpetologica*, 30: 156–60.

Hartman, B. 1978. "Description of major terrestrial and wetland habitats of Florida. In *Rare and Endangered Biota of Florida*, edited by R. W. McDiarmid, xvi–xix. Gainesville: University Presses of Florida.

Herrington, R. E. 1974. "Notes on a brood of *Coluber constrictor helvigularis* Auffenberg." *Herpetological Review*, 5: 38–39.

Highton, R. 1956. "Systematics and variation of the endemic Florida snake genus *Stilosoma*." *Bulletin of the Florida State Museum*, 1: 73–96.

Himes, J. G. 1998. *Activity patterns, habitat selection, excavation behavior, growth rates, and conservation of the Louisiana Pine Snake (Pituophis melanoleucus ruthveni)*. Unpubl. M.S. thesis. Shreveport: Louisiana State University. 58 pp.

Hoover, C. M. 2000. "The United States role in the international live reptile trade." *Amphibian and Reptile Conservation*, 2: 30–31.

Hutchinson, V. H. 1961. "Critical thermal maxima in salamanders." *Physiological Zoology*, 34: 92–125.

International Species Inventory System. 2005. Eagan, Minn.: ISIS.

Irwin, K. J., S. L. Collins, and J. T. Collins. 1993. "*Rhadinaea flavilata*: geographic distribution." *Herpetological Review*, 24: 110.

Irwin, L. K., and K. J. Irwin. 2002. "*Hyla femoralis*. geographic distribution." *Herpetological Review*, 33: 145.

Irwin, L. K., K. J. Irwin, T. W. Taggart, J. T. Collins, and S. L. Collins. 2001. "A herpetofaunal survey of the Apalachicola barrier islands, Florida: the 2000–2001 season." *Kansas Herpetological Society Newsletter*, 123: 12–15.

Irwin, K. J., T. E. Lewis, J. D. Kirk, S. L. Collins, and J. T. Collins. 2003. "Status of the Eastern Indigo Snake (*Drymarchon couperi*) on St. Vincent National Wildlife Refuge, Franklin County, Florida." *Journal of Kansas Herpetology*, 7: 13–20.

Jackson, D. R. 1991. "Multiple clutches and nesting behavior in the Gulf Coast Box Turtle." *Florida Field Naturalist*, 19: 14–16.

Jennings, R. D., and T. H. Fritts. 1983. "The status of the Black Pine Snake, *Pituophis melanoleucus lodingi* and the Louisiana Pine Snake, *Pituophis melanoleucus ruthveni*." Washington, D.C.: U.S. Fish and Wildlife Service and University of New Mexico Museum of Southwestern Biology.

Jensen, J. B. 1996. "*Heterodon simus* (Southern Hognose Snake) hatchling size." *Herpetological Review*, 27: 25.

Jobson, H. G. M. 1940. "Reptiles and amphibians from Georgetown County, South Carolina." *Herpetologica*, 2: 39–43.

Johnson, S. A. 2002. "Life history of the Striped Newt at a north-central Florida breeding pond." *Southeastern Naturalist*, 1: 381–402.

———. 2003. "Orientation and migration distances of a pond-breeding salamander (*Notophthalmus perstriatus*, Salamandridae). *Alytes*, 21: 3–22.

———. 2005. "Conservation and life history of the Striped Newt: the importance of habitat connectivity." In *Amphibians and Reptiles: Status and Conservation in Florida*, edited by W. E. Meshaka and K. J. Babbitt, 91–98. Malabar, Fla.: Krieger.

Knight, J. L. 1986. "Variation in snout morphology in the North American snake *Pituophis melanoleucus* (Serpentes: Colubridae)." *Journal of Herpetology*, 20: 77–79.

Krysko, K. L. 1995. *Resolution of the controversy regarding the taxonomy of the kingsnake,* Lampropeltis getula, *in southern Florida*. Unpublished M.S. thesis. Miami: Florida International University.

———. 2001. *Ecology, conservation, and morphological and molecular systematics of the kingsnake,* Lampropeltis getula *(Serpentes: Colubridae)*. Unpublished Ph.D. diss. Gainesville: University of Florida. x + 159 pp.

———. 2002. "Seasonal activity of the Florida Kingsnake *Lampropeltis getula floridana* (Serpentes: Colubridae) in southern Florida." *American Midland Naturalist*, 148: 102–14.

Krysko, K. L., and W. S. Judd. 2006. "Morphological systematics of kingsnakes, *Lampropeltis getula* complex (Serpentes: Colubridae), in the eastern United States. *Zootaxa*, 1193: 1–39.

LaClaire, L. V. 1999. "Final rule to list the Flatwoods Salamander as a threatened species." *Federal Register*, 50 CFR Part 17, 64: 15691–704.

Landers, J. L., and D. W. Speake. 1980. "Management and needs of sandhill reptiles in southern Georgia." *Proceedings of the Annual Conference of the Southeastern Association of Fish and Wildlife Agencies*, 34: 515–29.

Lawler, H. E. 1997. "The status of *Drymarchon corais couperi*, the Eastern Indigo Snake, in the southeastern United States." *Herpetological Review*, 8: 76–79.

Macey, J. R., T. J. Papenfuss, J. V. Kuehl, H. M. Fourcade, and J. L. Boore. 2004. "Phylogenetic relationships among amphisbaenian reptiles based on complete mitochondrial genomic sequences." *Molecular Phylogenetics and Evolution* 33: 22–31.

Means, D. B., and K. L. Krysko. 2001. "Biogeography and pattern variation of kingsnakes, *Lampropeltis getula*, in the Apalachicola region of Florida." *Contemporary Herpetology* 2001: 1–7.

Means, D. B., and C. J. Longden. 1976. "Aspects of the biology and zoogeography of the Pine Barrens Treefrog (*Hyla andersoni*) in northern Florida." *Herpetologica*, 32: 117–30.

Means, D. B., J. G. Palis, and M. Baggett. 1996. "Effects of slash pine silviculture on a Florida population of Flatwoods Salamander." *Conservation Biology*, 10: 426–37.

Mecham, J. S. 1967. "*Notophthalmus perstriatus* (Bishop). Striped Newt." *Catalogue of American Amphibians and Reptiles*, 38: 1–2.

Mecham, J. S., and R. E. Hellman. 1952. "Notes on the larvae of two Florida salamanders." *Quarterly Journal of the Florida Academy of Science*, 15: 127–33.

Meylan, P. A. 1984. "The northwestern limit of distribution of *Rhineura floridana* with comments on the dispersal of amphisbaenians." *Herpetological Review*, 15: 23–24.

Minx, P. 1992. "Variation in the phalangeal formulae in the turtle genus *Terrapene*." *Journal of Herpetology*, 26: 234–38.

Morgareidge, K. R., and H. T. Hammel. 1975. "Evaporative water loss in box turtles: Effects of rostral brainstem and other temperatures." *Science*, 187: 366–68.

Morin, P. J. 1983. "Competitive and predatory interactions in natural and experimental populations of *Notophthalmus viridescens* and *Ambystoma tigrinum*." *Copeia*, 1983: 628–39.

Mosimann, J. E., and G. B. Rabb. 1948. "The salamander *Ambystoma mabeei* in South Carolina." *Copeia*, 1948: 304.

Mount, R. H. 1963. "The natural history of the red-tailed skink, *Eumeces egregius* Baird." *American Midland Naturalist*, 70: 356–85.

———. 1965. "Variation and systematics of the scincoid lizard *Eumeces egregius* (Baird)." *Bulletin of the Florida State Museum*, 9: 183–213.

———. 1968. "*Eumeces egregius* (Baird). Mole Skink." *Catalogue of American Amphibians and Reptiles*. 73.1–73.2.

———. 1975. *The Reptiles and Amphibians of Alabama*. Tuscaloosa: University of Alabama Press. 347 pp.

Mushinsky, H. R. 1984. "Observations of the feeding habits of the Short-tailed Snake, *Stilosoma extenuatum*, in captivity." *Herpetological Review*, 15: 67–68.

Mushinsky, H. R., and B. W. Witz. 1993. "Notes on the Peninsula Crowned Snake, *Tantilla relicta*, in periodically burned habitat." *Journal of Herpetology*, 27: 468–70.

Myers, C. W. 1967. "The Pine Woods Snake, *Rhadinaea flavilata* (Cope)." *Bulletin of the Florida State Museum*, 11: 47–97.

NatureServe. 2005. "*Rhadinaea flavilata*." *NatureServe Explorer: an online encyclopedia of life* [Web application]. Version 4.5. Arlington, Va.: NatureServe. <http://www.natureserve.org/explorer>, accessed November 1, 2005.

Neill, W. T. 1948. "Hibernation of amphibians and reptiles in Richmond County, Georgia." *Herpetologica*, 4: 107–14.

———. 1951a. "Notes on the natural history of certain North American snakes." *Publication of the Research Division of Ross Allen's Reptile Institute* 1: 47–60.

———. 1951b. "The eyes of the worm lizard, and notes on the habits of the species." *Copeia*, 1951: 177–78.

———. 1954. "Evidence of venom in snakes of the genera *Alsophis* and *Rhadinaea*." *Copeia*, 1954: 59–60.

———. 1957. "Distributional notes of Georgia's amphibians, and some corrections." *Copeia*, 1957: 43–47.

Neill, W. T., and R. Allen. 1949. "A new kingsnake (genus *Lampropeltis*) from Florida." *Herpetologica*, 5: 101–6.

Nelson, D. H., J. D. Cochran, C. G. Drew, and T. D. Schwaner. 1994. "*Rhadinaea flavilata*: geographic distribution." *Herpetological Review*, 25: 34.

Noble, G. K., and R. C. Noble. 1923. "The Anderson Treefrog (*Hyla andersoni* Baird). Observations on its habits and life history." *Zoologica*, 2(18): 416–52.

Outcalt, K. W., and P. A. Outcalt. 1994. "The longleaf pine ecosystem: an assessment of current conditions." Unpublished report on USDA Forest Service Forest Inventory and Analysis Data. 23 pp.

Palis, J. G. 1992. "Geographic distribution: *Rhadinaea flavilata*." *Herpetological Review*, 23: 92.

———. 1997a. "Breeding migration of *Ambystoma cingulatum* in Florida." *Journal of Herpetology*, 31: 71–78.

———. 1997b. "Distribution, habitat, and status of the Flatwoods Salamander (*Ambystoma cingulatum*) in Florida, USA." *Herpetological Natural History*, 5(1): 53–65.

Palmer, W. M. 1987. "A new species of glass lizard (Anguidae: *Ophisaurus*) from the southeastern United States." *Herpetologica*, 43: 415–23.

Paulson, D. R. 1966. "Variation in some snakes from the Florida Keys." *Quarterly Journal of the Florida Academy of Science*, 29: 295–308.

Pembleton, E. F., and S. L. Williams. 1978. "*Geomys pinetis*." *Mammalian Species*, 86: 1–3.

Petranka, J. W. 1998. *Salamanders of the United States and Canada*. Washington and London: Smithsonian Institution Press.

Pope, C. H. 1950. "A statistical and ecological study of the salamander *Plethodon yonahlossee*." *Bulletin of the Chicago Academy of Science*, 9: 79–106.

Porras, L., and L. D. Wilson. 1979. "New distributional records for *Tantilla oolitica* Telford (Reptilia, Serpentes, Colubridae) from the Florida Keys." *Journal of Herpetology*, 13: 218–20.

Price, W. H., and L. G. Carr. 1943. "Eggs of *Heterodon simus*." *Copeia*, 1943: 193.

Reed, R. M., L. D. Vorhees, and P. J. Mulholland. 1981. "Environmental impacts associated with using peat for energy." *Reprints of the 1981 International Gas Research Conference*, U.S. Department of Energy, 1048–57.

Reichling, S. B. 1986. "The endangered pine snakes of the gulf coast; their ecology and husbandry." *AAZPA Regional Conference Proceedings*, 119–25.

———. 1988. "Reproduction in captive Louisiana Pine Snakes, *Pituophis melanoleucus ruthveni*." *Herpetological Review*, 19: 77–78.

———. 1990. "Reproductive traits of the Louisiana Pine Snake, *Pituophis melanoleucus ruthveni* (Serpentes: Colubridae)." *Southwestern Naturalist*, 35: 221–22.

———. 1992. *North American Regional Studbook for the Black Pine Snake*, Pituophis melanoleucus lodingi, *and the Louisiana Pine Snake*, Pituophis melanoleucus ruthveni. 1st ed. Memphis, Tenn.: The Memphis Zoo.

———. 1995. "The taxonomic status of the Louisiana Pine Snake (*Pituophis melanoleucus ruthveni*) and its relevance to the evolutionary species concept." *Journal of Herpetology*, 29: 186–98.

———. 2000. *North American Regional Studbook for the Louisiana Pine Snake*, Pituophis ruthveni. Memphis, Tenn.: The Memphis Zoo.

Reichling, S. B., and P. Louton. 1989. "Geographic distribution: *Rhadinaea flavilata*." *Herpetological Review*, 20: 76.

Richter, S. C., and R. A. Seigel. 2002. "Annual variation in the population ecology of the endangered gopher frog, *Rana sevosa* Goin and Netting." *Copeia*, 2002: 962–72.

Richter, S. C., J. E. Young, G. N. Johnson, and R. A. Seigel. 2003. "Stochastic variation in reproductive success of a rare frog, *Rana sevosa*: implications for conservation and for monitoring amphibian populations." *Biological Conservation*, 111: 171–77.

Richter, S. C., J. E. Young, R. A. Seigel, and G. N. Johnson. 2001. "Postbreeding movements of the dark gopher frog, *Rana sevosa*, Goin and Netting: implications for conservation management." *Journal of Herpetology*, 35: 316–21.

Rossi, J., and R. Rossi. 1992. "Notes on the natural history, husbandry, and breeding of the Southern Hognose Snake (*Heterodon simus*)." *Vivarium*, 3: 16–18, 27.

———. 1993. "Notes on the captive maintenance and feeding behavior of a juvenile Short-tailed Snake (*Stilosoma extenuatum*)." *Herpetological Review*, 24: 100–101.

Rudolph, D. C., and S. J. Burgdorf. 1997. "Timber Rattlesnakes and Louisiana Pine Snakes of the west gulf coastal plain: hypothesis of decline." *Texas Journal of Science*, 49: 111–21.

Rudolph, D. C., S. J. Burgdorf, R. N. Conner, C. S. Collins, C. M. Duran, M. Ealy, J. G. Himes, D. Saenz, R. R. Schaefer, and T. Trees. 2002. "Prey handling and diet of Louisiana Pine Snakes (*Pituophis ruthveni*) and Black Pine Snakes (*P. melanoleucus lodingi*) with comparisons to other selected colubrid snakes." *Herpetological Natural History*, 9: 57–62.

Rudolph, D. C., S. J. Burgdorf, R. R. Schaefer, R. N. Conner, and R. W. Maxey. 2006. "Status of *Pituophis ruthveni* (Louisiana Pine Snake)." *Southeastern Naturalist*, 5: 463–72.

Russell, K. R., H. G. Hanlin, and J. W. Gibbons. 1998. "*Ambystoma mabeei*. Breeding migration. Natural history notes." *Herpetological Review*, 29: 36–37.

Schwartz, A., and R. Etheridge. 1954. "New and additional herpetological records from the North Carolina coastal plain." *Herpetologica*, 10: 167–71.

Semlitsch, R. D. 2002. "Critical elements for biologically based recovery plans of aquatic-breeding amphibians." *Conservation Biology*, 16(3): 619–29.

Semlitsch, R. D., D. E. Scott, J. H. K. Pechmann, and J. W. Gibbons. 1996. "Structure and dynamics of an amphibian community: evidence from a 16-year study of a natural pond." In *Long-Term Studies of Vertebrate Communities*, edited by M. L. Cody and J. A. Smallwood, 217–48. San Diego: Academic.

Sharitz, R. R., and J. W. Gibbons. 1982. "The ecology of southeastern shrub bogs (pocosins) and Carolina bays: a community profile." FWS/OBS-82/04. Washington, D.C.: U.S. Department of the Interior, Fish and Wildlife Service, Division of Biological Services. 93 pp.

Slavens, F. 1988. *Inventory, Longevity, and Breeding Notes—Reptiles and Amphibians in Captivity*. Seattle: Frank L. Slavens. 401 pp.

Smith, C. R. 1982. "Food resource partitioning of Florida fossorial reptiles." In *Herpetological Communities*, edited by N. J. Scott, Jr., 173–78. Wildlife Research Report 13, U.S. Department of Interior, Fish and Wildlife Service.

Smith, H. M., D. Chizar, J. R. Staley II, and K. Trepedelen. 1994. "Populational relationships in the cornsnake *Elaphe guttata* (Reptilia: Serpentes)." *Texas Journal of Science*, 46: 259–92.

Smith, H. M., and J. P. Kennedy. 1951. "*Pituophis melanoleucus ruthveni* in eastern Texas and its bearing on the status of *P. catenifer*." *Herpetologica*, 7: 93–96.

Stejneger, L. 1894. "*Stilosoma extenuatum*." In "Notes on the reptiles and batrachians collected in Florida in 1892 and 1893." E. Loennberg, 323–24, *Proceedings of the U.S. National Museum*, 17: 317–39.

Sulentich, J. M., L. R. Williams, and G. N. Cameron. 1991. "*Geomys breviceps*." *Mammalian Species*, 383: 1–4.

Telford, S. R., Jr. 1955. "Notes on an exceptionally large worm lizard." *Copeia*, 1955: 258–59.

———. 1966. "Variation among the southeastern Crowned Snakes, genus *Tantilla*." *Bulletin of the Florida State Museum, Biological Sciences*, 10: 261–304.

Tennant, A. 1985. *A Field Guide to Texas Snakes*. Austin: Texas Monthly Press. 260 pp.

Thomas, R. A., B. J. Davis, and M. R. Culbertson. 1976. "Notes on variation and range of the Louisiana Pine Snake, *Pituophis melanoleucus ruthveni* Stull (Reptilia, Serpentes, Colubridae)." *Journal of Herpetology*, 10: 252–54.

Tuberville, T. D., J. R. Bodie, J. B. Jensen, L. LaClaire, and J. W. Gibbons. 2000. "Apparent decline of the Southern Hognose Snake, *Heterodon simus*." *Journal of the Elisha Mitchell Society*, 116: 19–40.

Tucker, J. K., R. S. Funk, and G. L. Paukstis. 1978. "The adaptive significance of egg morphology in two turtles (*Chrysemys picta* and *Terrapene carolina*). *Bulletin of the Maryland Herpetological Society*, 14: 10–22.

Turner, R. L. 2003. "*Rhadinaea flavilata*: geographic distribution." *Herpetological Review*, 34: 390.

U.S. Fish and Wildlife Service. 1999. "Bluetail Mole Skink, *Eumeces egregius lividus*." In *South Florida Multi-Species Recovery Program. A Species Plan. . . an Ecosystem Approach*, 4.529–540. Atlanta: U.S. Fish and Wildlife Service, Southeast Region.

Vandeventer, T. L., and R. A. Young. 1989. "Rarities of the longleaf: the Black and Louisiana Pine Snakes." *Vivarium*, 1: 32–36.

Weaver, W. G., and S. P. Christman. 1978. "Big Pine Key Ringneck snake." In *Rare and Endangered Biota of Florida*. Vol. 3. *Amphibians and Reptiles*, edited by Paul Moler, 41–42. Gainesville: University Presses of Florida.

Whiteman, H. H., T. M. Mills, D. E. Scott, and J. W. Gibbons. 1995. "Confirmation of a range extension for the Pine Woods Snake (*Rhadinaea flavilata*)." *SREL Reprint*, 1989: 1–2.

Wiley, J. E. 2003. "Replication banding and FISH analysis reveal the origin of the *Hyla femoralis* karotype and XY/XX sex chromosomes." *Cytogenic and Genome Research*, 101: 80–83.

Wilks, B. J. 1962. "The pine snake in central Texas." *Herpetologica*, 18: 108–10.

Williams, A. A., and J. E. Cordes. 1996. "Geographic distribution: *Pituophis ruthveni*." *Herpetological Review*, 27: 35.

Wilson, D. S., Mushinsky, H. R., and R. A. Fischer. 1997. "Species profile: Gopher Tortoise (*Gopherus polyphemus*) on military institutions in the southeastern United States." Technical Report SERDP-97-10. Vicksburg, Miss.: U.S. Army Engineer Waterways Experiment Station.

Wilson, L. D. 1970. "The racer *Coluber constrictor* (Serpentes: Colubridae) in Louisiana and eastern Texas." *The Texas Journal of Science*, 22 (September): 67–85.

Wood, K. N. 1998. "*Rhadinaea flavilata*: geographic distribution." *Herpetological Review*, 29: 249.

Woolfenden, G. E. 1962. "A range extension and subspecific relations of the Short-tailed Snake, *Stilosoma extenuatum*." *Copeia*, 1962: 648–49.

Wright, A. H. 1935a. "Some rare amphibians and reptiles of the United States (Abstract)." *Science*, 81: 463.

———. 1935b. "Some rare amphibians and reptiles of the United States." *Proceedings of the National Academy of Sciences, U.S.*, 21: 340–45.

Wright, A. H., and A. A. Wright. 1949. *Handbook of Frogs and Toads of the United States and Canada*. Ithaca, N.Y.: Comstock.

———. 1957. *Handbook of Snakes*. Vol 1. Ithaca, N.Y.: Comstock.

Young. D. P., Jr. 1988. "*Rhadinaea flavilata*. geographic distribution." *Herpetological Review*, 19: 20.

Young, R. A., and T. L. Vandeventer. 1988. "Recent observations on the Louisiana Pine Snake, *Pituophis melanoleucus ruthveni* (Stull)." *Bulletin of the Chicago Herpetological Society*, 23: 203–7.

Index

Steve Reichling is curator of reptiles, aquatics, and small mammals at the Memphis Zoo, where he has worked for 30 years. He is also the author of *Tarantulas of Belize* (2003).